# Ed Shaffer

# Canada's Oil and the American Empire

AF691209

Hurtig Publishers
Edmonton

*To Florence, Martha, and Paul*

Copyright © 1983 by Ed Shaffer.

No part of this book may be reproduced or transmitted in any form by any means, electronic or mechanical, including photocopying and recording, or by any information storage or retrieval system, without written permission from the publisher, except for brief passages quoted by a reviewer in a newspaper or magazine.

Hurtig Publishers Ltd.
10560 - 105 Street
Edmonton, Alberta

**Canadian Cataloguing in Publication Data**

Shaffer, Edward H.
Canada's oil and the American empire

Includes index.
Bibliography: p.
ISBN 0-88830-233-9 (bound).—ISBN
0-88830-234-7 (pbk.)

1. Petroleum industry and trade—Government policy—United States. 2. Petroleum industry and trade—Political aspects—United States. 3. Petroleum industry and trade—Political aspects—Canada. I. Title.

HD9566.S496 338.2′7282′0973 C83-091072-7

Edited by José Druker
Designed by David Shaw & Associates Ltd.
Composed by Attic Typesetting Inc.
Printed and bound in Canada by D.W. Friesen & Sons Ltd.

# Contents

# List of Tables

# Preface

This book is an enlarged and updated version of another recent book, *The United States and the Control of World Oil*, published by Croom Helm in England and St. Martin's Press in the United States. As can be gleaned from the title, the earlier work was concerned with worldwide United States oil policies. Though it contained sections dealing with Canadian oil questions, it was not primarily concerned with the relationship between the United States and Canada.

The present book, while leaving the earlier work essentially intact, has paid much more attention to the role of Canada. Significant data on Canada have been inserted throughout. In addition, two chapters dealing specifically with Canadian problems have been added. These additions should enable the reader to acquire a better understanding of Canada–U.S. relations within the world oil picture.

In writing this book I have received suggestions, aids, and comments from many people, too many to mention in a brief acknowledgement. Nevertheless, I would like to express my thanks to several individuals who were particularly helpful. Dr. Larry Pratt, professor of Political Science at the University of Alberta, gave me very valuable assistance in writing the Canadian sections. He also directed me to some important sources of information for the book as a whole, as did David Leadbeater, presently a graduate student at the University of Toronto. Dr. Gilbert Reschenthaler, Faculty of Business Administration and Commerce, University of Alberta, made many valuable comments on the beginning chapters. I want to render special thanks to Dr. Clement Leibovitz, Computing Services, University of Alberta, for the many hours he spent in helping prepare the manuscript for the computer. Finally, I would like to thank Energy, Mines and Resources Canada for the financial assistance I received to prepare this book. Needless to say, the opinions expressed are mine entirely and do not necessarily reflect the opinions of those who helped me.

CHAPTER ONE

# Introduction: The Age of Oil

There is a struggle raging in the world, a struggle over energy. During the 1970s the industrialized countries faced the spectre of an oil famine, a spectre which they seemed unable to exorcise. Though this spectre has receded in the early 1980s, it is sure to rise again and will continue to haunt the world for many years.

The events of the 1970s have shaken the very foundations of the major industrialized market societies. In those years the world was transformed from one in which oil was both cheap and readily available to one in which it became dear and relatively scarce.

The new era is forcing these societies to make many painful adjustments, of both a physical and a social nature. The physical involves finding more efficient ways to use oil and developing substitutes for it. The social implies major shifts in the distribution of world income and concomitant changes in world power relationships.

Of the two adjustments, the social is undoubtedly the more complex. In comparison, the physical is relatively simple. Science and technology will find ways to use oil more efficiently as well as to discover substitutes for it. However, they will not be able to deal with the social disturbances that will accompany the end of the oil era.

The task of predicting, analysing, and offering possible solutions to these upheavals should fall to the social scientists. Unfortunately, today's social scientists, and particularly economists, are not equipped to handle this task. Economists have abandoned the tradition of political economy in favour of a more abstract, allegedly more value-free "positive" economics. However, the concept of a value-free science is an absurdity: all scientists are products of the value systems of the societies in which they live, and the choice of research paths in both the physical and social sciences always reflects these systems. More important is that the pursuit of a value-free science has caused economists to concentrate on testing narrowly defined hypotheses. The advent of the computer has further led them to test only those hypotheses that are readily quantifiable.[1] In doing so they have avoided raising basic questions about the social structures of modern societies. They have accepted these structures as given.

Though economists like Adam Smith, Karl Marx, and Joseph Schumpeter recognized relationships between technical and social change, modern economists have by and large ignored these relationships. It is not surprising, therefore, that their analyses of the world's economic behaviour in the 1970s were, as in the 1930s, way off the mark. As Wassily Leontieff remarked in his presidential address to the American Economic Association, "In no other field of empirical inquiry has so massive and sophisticated a statistical machinery been used with such indifferent results."[2]

For this reason, this book proposes to examine the problems posed by the rise and fall of oil in the tradition of political economy. It seeks to analyse, in particular, the relations between technical and social change as they apply to oil. Only through an understanding of these relationships can one comprehend the implications of the rise and fall of oil.

In Schumpeterian terms the ushering in of the oil era was an example of "the perennial gale of creative destruction."[3] The creative winds of the new product, oil, destroyed the old one, coal. But this "perennial gale" does something more than replace an old product with a new one. It also displaces the established entrepreneurs (*i.e.*, those making the old product) with a new group of entrepreneurs (*i.e.*, those introducing the new product).[4] The rise of this new group brings a change in the power relationships of the economy. Furthermore, on an international scale, it may bring a change in the power relationships among nations. These changes in international power relationships did not concern Schumpeter. They are, however, central to the theme of this book.

Its theme is that the rise of oil in the United States gave birth to a new group of entrepreneurs, the international oil companies, who began to play an increasingly important role in the domestic economy. Furthermore, the very nature of oil—a commodity whose supply is finite and whose geographic distribution is highly unequal—has led these companies to expand into other countries in their search for oil. Their expansion could have not been successful without the active help of the U.S. government, which pursued a deliberate policy of acquiring domination over the world's oil. The rise of oil hence was associated with the rise of the American Empire and with the attempts by the U.S. to establish hegemony over the rest of the world.

New forces now at work are undermining America's position. The U.S. is no longer able to control oil in many formerly "secure" areas. Not only is this inability evident in the nations belonging to the Organization of Petroleum Exporting Countries (OPEC), but it is also evident in Canada, where a struggle is going on to free its resources from U.S. control.

In addition, the end of the oil era may further weaken the U.S. position. There is no guarantee that the United States will be able to control the new energy sources and the new technology necessary to make them operable. To retain its dominant position the U.S. will be forced to attempt to do so, and all the world's industrialized countries, and especially the market economies, will be affected as they engage in a race with the U.S. to be the first to develop the new technologies. One aspect of this struggle to come is the introduction in all the major market economies of austerity programs, which are designed to shift a greater share of national income from consumption into investment in energy projects.

This book intends to analyse this struggle, this new form of economic warfare. It will do so by tracing (1) the rise of oil in the United States, (2) the ascendancy of the U.S. to a position of world dominance, (3) the impact of U.S. oil policies on Canada, and (4) the erosion of the U.S. world position.

## Notes

1. Economists often try to take account of qualitative variables through the use of dummy variables. While useful in some types of research, this technique usually can give only a superficial understanding of the relationships between quantitative and qualitative variables.
2. W. Leontieff, "Theoretical Assumptions and Unobserved Facts," *American Economic Review*, vol. 61, no. 1 (1971), p. 3. For other criticisms of modern economics see J. Robinson, "The Second Crisis of Economic Theory," *American Economic Review*, vol. 62, no. 2 (1972), pp. 1–10, and J. K. Galbraith, "Power and the Useful Economist," *American Economic Review*, vol. 63, no. 1 (1973), pp. 1–11.
3. J. Schumpeter, *Capitalism, Socialism, and Democracy*, 2nd ed. (Harper and Brothers, New York and London, 1947), p. 84.
4. Schumpeter's thesis that new entrepreneurs are associated with new products is only partially true. While it does portray the pattern in oil, it is inapplicable in other industries. Innovations have also been introduced by established firms, *e.g.*, nylon by DuPont and the transistor by Bell Labs.

CHAPTER TWO

# The Early Expansion of American Capitalism

From its birth in 1776 the United States has pursued expansionist policies. This is not surprising since the American Revolution occurred during the Mercantilist Era, an era in which all the major powers were also pursuing such policies. While the Revolution has often been interpreted as a revolt against mercantilism in general, it was in actuality a revolt against the application of specific British mercantilist policies to the North American colonies.

One of these policies was the attempt by the British to reserve the western frontier areas for the fur trade. This clashed with the desires of settlers and land speculators to bring these lands under their control. It was thus to be expected that, with the ouster of the British, the fledgling American state would emulate its European rivals in a policy of expansion.

The aggressiveness of the young republic can be shown through the relatively rapid growth of territory under its control. In 1783, the year in which Britain signed the peace treaty with the U.S., its area consisted of approximately 500,000 square miles. By 1853, seventy years later, the country extended "from sea to shining sea," encompassing an area of over three million square miles.[1] First the central part of the North American continent was acquired, then the nation's interests turned outward.

## Continental Consolidation

The conquest of the continent was not a peaceful one. It required three major wars (the War of 1812, the Mexican War, and the Civil War) and a host of minor ones against the Indians. The latter, though not on the scale of the major conflicts, were nevertheless bloody and costly.[2] All these wars had the effect of spreading capitalism across the continent, thereby creating the basis for the expansion of the U.S. industrial machine.

The war of 1812 secured the boundaries of the young republic against reconquest by England. The Mexican and Indian wars enabled the U.S. to conquer the West. As a result of the Mexican War the nation added more than 800,000 square miles to its territory.[3] This territory includes, among others, the states of Texas, California, New Mexico, Arizona, Nevada, and Colorado, containing some of the richest mineral deposits in the world. Two of these states, Texas and California, have been instrumental in enabling the United States to establish itself as a major oil power.

The Civil War introduced capitalism into the South. The North's victory enabled its industrialists to expand their operations in that region. As the freed slaves entered the market economy, they provided both a market for manufactured goods and a relatively cheap source of labour. The growth of a market in the South encouraged the building of railroads, further integrating the South into the national economy. The *ante-bellum* South, it should be noted, had stronger economic links with England than with the North, because of which Britain had supported the Confederacy during the Civil War. The weakening of these links gave an important impetus to the expansion of northern industry.

The decimation of the Indians in the Indian wars[4] opened the vast western territory to settlement and exploitation. With settlement and exploitation came the transcontinental railroads, which linked these areas to the industrial East. The railroads, however, did more than merely provide economic links between East and West. They provided the basis for a major restructuring of the U.S. economy.

By integrating the economies of the South and West with the rapidly industrializing North-East, they created one national market. A unified common market displaced a series of disparate regional ones. The new national market gave birth to national firms, which either bought out or destroyed the small-scale regional enterprises that had characterized the pre–Civil War economy. The large company became the symbol of American capitalism.

In the period between the end of the Civil War and World War I, the United States witnessed the greatest consolidation movement in its history. These consolidations took the forms of trusts, holding companies, and mergers.[5] This period gave birth to such giant firms as United States Steel, American Tobacco, International Harvester, Anaconda, American Sugar Refining, Armour, Standard Oil of New Jersey, the large railroads, and many other giant corporations.[6] All these companies acquired many plants, mines, and establishments, which enabled them to

**Table 2.1: U.S. Plants Controlled by Major Corporations, 1901**

| *Corporation* | *Plants* (Number) |
|---|---|
| Amalgamated Copper Co. | 11 |
| American Smelting and Refining Co. | 121 |
| American Sugar Refining Co. | 55 |
| Consolidated Tobacco Co. | 150 |
| Standard Oil Co. of New Jersey | 400 |
| United States Steel Corp. | 785 |

Source: J. Moody, *The Truth about the Trusts* (Moody Publishing, Chicago, 1904), p. 453.

achieve dominant positions both in their respective markets and in the economy as a whole. Table 2.1 lists the number of plants controlled by some of the major firms at the turn of the century.

This form of growth, *i.e.*, expansion through the acquisition of control of production facilities in diverse geographic regions, represented "the wave of the future." It was inevitable that this growth would breach the confines of the domestic economy and spill over into foreign ones. With the consolidation of the central portion of the continent completed, the American state turned its eyes towards other areas.

## External Expansion

Shortly after the end of the Civil War in 1865, the United States began to consider seriously expansion beyond the mid-continent. It made its first important moves in 1867, when it annexed the Midway Islands in the Pacific and purchased Alaska from Russia.

The latter move, in particular, had important implications for Canada. It was seen both in Canada and the United States as a move to forestall Confederation and to occupy Canada. Ryerson quotes U.S. Senator Charles Sumner of Massachusetts as stating:

> The present treaty [with Russia for the purchase of Alaska] is a visible step in the occupation of the whole American continent. As such it will be recognized by the world and accepted by the American people.[7]

The Alaskan purchase occurred at a time when there was an active campaign in the United States to annex its neighbour to the north. In 1866, according to Ryerson, a bill was introduced in the U.S. Congress entitled "A Bill for the Admission of the States of Nova Scotia, New

Brunswick, Canada East and Canada West, and for the organization of the Territories of Selkirk, Saskatchewan and Columbia" to be "admitted as States and Territories of the United States of America."[8]

Though this bill never passed, the "machinations"—in the words of Ryerson—of the annexationists did not cease.[9] The U.S. government opposed Canadian Confederation, and the U.S. House of Representatives passed a resolution expressing "extreme solicitude" over it.[10] This attitude prompted Sir John A. Macdonald to write:

> It is quite evident to me. . . that the U.S. government are resolved to do all they can, short of war, to get possession of the Western territory, and we must make immediate and vigorous steps to counteract them.[11]

This threat to Canada was instrumental in spurring Confederation. As Ryerson observed:

> The danger of a military collision with the U.S. loomed large in the thinking of Canadian and Maritime leaders. It proved decisive in bringing London around to agree to the Confederation proposals.[12]

Blocked by Confederation, the United States concentrated on expansion beyond the territorial confines of the continent.

Its next major step was the Spanish–American War of 1898. As a result of that war the U.S. acquired Puerto Rico, the Philippines, Guam, Hawaii, and Samoa. The first three were seized from Spain. Hawaii, which had been nominally independent at the outset of the war, was formally incorporated in the United States while Samoa, which had had a protectorate status at the outbreak of hostilities, was ceded to the United States by Germany and Great Britain, who also had claims to that island chain. In the years following the Spanish–American War the U.S. acquired control of the Panama Canal Zone and the Virgin Islands, and obtained a lease on the Corn Islands from Nicaragua. These acquisitions are shown in Table 2.2.

Despite these acquisitions the United States possessed a relatively small empire. Its overseas territories contained only 700,000 square miles, of which almost 600,000 were in the vast emptiness of Alaska. In comparison with the empires of England and France, its overseas territory was relatively insignificant. Nevertheless, the United States became a major world power during this period because it was able to exercise hegemony over most of the countries in the Western Hemisphere without formally incorporating them in the Union. This policy of rule without incorporation was best exemplified by the treatment of Cuba.

**Table 2.2: Overseas Territorial Expansion of the United States, 1867–1917**

| *Territory* | *Year Acquired* | *Area* (Thousand Sq. Miles) | *Method of Acquisition* |
|---|---|---|---|
| Alaska | 1867 | 586 | Purchase from Russia. |
| Midway Islands | 1867 | a | Formally annexed. First reported and claimed for U.S. in 1859 by Navy Capt. N. C. Brooks. |
| Hawaii | 1898 | 6 | Independent kingdom seized by U.S. |
| The Philippines | 1898 | 116 | From Spain as a result of Spanish-American War. |
| Puerto Rico | 1899 | 3 | From Spain as a result of Spanish-American War. |
| Guam | 1899 | a | From Spain as a result of Spanish-American War. |
| Samoa | 1900 | a | From treaty with England and Germany. |
| Panama Canal Zone | 1904 | 1 | From treaty imposed on Panama. |
| Corn Islands | 1914 | a | Leased from Nicaragua for ninety-nine years in treaty imposed on that country. |
| Virgin Islands | 1917 | a | Purchased from Denmark. |
| Total | | 712 | |

a. Less than 500 square miles.

Sources: *Encyclopaedia Britannica* (Chicago, 1962), vol. 1, p. 501, and vol. 15, p. 460; *Historical Statistics of the United States, Colonial Times to 1957: A Statistical Abstract Supplement* (U.S. Government Printing Office, Washington, 1960), p. 236; S. Nearing & J. Freeman, *Dollar Diplomacy*, reprint (Monthly Review Press, New York and London, 1966), p. 283.

*Acquisition Without Annexation*

Instead of annexing Cuba, the United States gave that country formal independence. At the same time it forced the Cubans to incorporate the Platt Amendment in their constitution. This amendment, which was introduced into the U.S. Congress by Senator Platt of Connecticut, granted the United States "the right to intervene for the preservation of Cuban independence, the maintenance of a government adequate for the protection of life, property, and individual liberty" and stipulated that "the Government of Cuba will sell or lease to the United States lands necessary for coaling or naval stations at certain specified points, to be agreed upon with the President of the United States."[13] In 1903 the United States used this latter stipulation to obtain a naval base at Guatanamo,[14] which it has occupied ever since.

There were a number of reasons why the U.S. decided not to annex Cuba. First, it did not want the responsibility for caring for the Cubans. Second, it felt it could obtain the results it wanted, *i.e.*, economic domination, without going through the costly process of formal colonization. In fact during the Spanish–American War a debate raged over the necessity of incorporating any new territories in the Union. Not only was there opposition to incorporating Cuba but to incorporating Hawaii and the Philippines as well. The latter two were incorporated only because they were regarded as essential stepping stones to the potentially lucrative trade with China. Their incorporation is, however, to be perceived as the exception to the general rule of non-annexation.

This attitude towards formal annexation was perhaps best expressed by Carl Schurz, one of the country's most influential political leaders and a man who probably would have been president had he not been born in Germany.[15] In a famous article, "Manifest Destiny," written for *Harper's* in 1893, he argued against colonialism of any kind, stating that the U.S. could gain its economic objectives "without taking those countries into our national household . . . without assuming any responsibilities for them."[16] The secretary of state at that time, Walter Gresham, commented that this was "the best article of the kind I have ever seen."[17] Several years later, at an 1898 meeting of the National Civic Federation, Schurz spelled out more clearly what he meant. In commenting on that speech W. A. Williams notes:

> Schurz . . . had always accepted the necessity of market expansion while opposing traditional colonialism. Exports, he had long argued, "will save us from the suffocation so much dreaded. . . . I fully agree . . . we cannot have too many." But the United States could obtain the required bases and ports "without burdening itself with any political responsibilities in the regions concerned." Hence the proper policy was to occupy the outposts "until they are thoroughly pacified"; and then, having obtained the required economic and military footholds, withdraw. That would extend freedom by "exerting civilizing influences upon the population of the conquered territories," and "gain commercial opportunities of so great a value that they will more than compensate for the cost of the war."[18]

### *Dollar Diplomacy*

The policy of using American military might more for market expansion and commercial gain than for empire building in the traditional sense was reinforced by adoption of the "open door" policy towards China. Under that policy, the U.S. demanded that the Great Powers allow it equal access to the markets of China. The U.S. also engaged in numerous landings in and temporary occupations of Latin American countries to ensure the protection of its economic interests.[19] It pursued a policy of "dollar diplomacy," whose essence was summarized in a 1912 speech by President Taft:

> The diplomacy of the present administration has sought to respond to modern ideas of commercial intercourse. This policy has been characterized as substituting dollars for bullets. . . . It is an effort frankly directed to the increase of American trade upon the axiomatic principle that the Government of the United States shall extend all proper support to every legitimate and beneficial American enterprise abroad. . . . If this Government is really to preserve to the American people that free opportunity in foreign markets which will soon be indispensable to our prosperity, even greater effort must be made. Otherwise the American merchant, manufacturer, and exporter will find many a field in which American trade should logically predominate preempted through the more energetic efforts of other governments and other commercial nations. . . . We need legislation enabling the members of the foreign service to be systematically brought into direct contact with the industrial, manufacturing, and exporting interests of this country in order that American business men may enter the foreign field with a clear perception of the exact conditions to be dealt with and the officers themselves may prosecute their work with a clear idea of what American industrial and manufacturing interests require.[20]

While Taft's address clearly reflected the ideas of Carl Schurz, its emphasis was somewhat different. Schurz was mainly concerned with using U.S. power as a means of opening up foreign markets for U.S. goods. Taft extended this concept to the protection of "American enterprises abroad." This change in emphasis was important because, by the time Taft had assumed the presidency, the export of capital had become increasingly important for the U.S. economy.

During the nineteenth century the United States was a net importer of capital. It had to export goods to pay for these capital imports. It was for this reason that American foreign policy was initially directed at preserving and expanding markets for American exports. However, after continental consolidation had been achieved, and especially after the large national corporations had emerged, the U.S. began to export capital as well as goods. The structural changes within the domestic economy led to changes in the composition of exports, with the export of capital becoming prominent. This shift to the export of capital, a shift that has occurred in all the major industrial market economies, has been described by both Hobson and Lenin as one of the essential features of modern imperialism.[21]

The shift to significant capital exports probably occurred in the U.S. during the 1890s. In that period net capital flows changed from a positive to a negative item in the balance-of-payments account. In 1895, for instance, the U.S. recorded a net capital inflow of $137 million. This inflow was reduced to $40 million in 1896 and in 1897 changed to a net outflow of $23 million. This net outflow jumped sharply to $279 million during 1897 and remained well above $200 million for the rest of the decade.[22]

Not all this net outflow represented U.S. investment overseas. Prior to 1900 the capital account was used as a balancing item, *i.e.*, errors and omissions were placed in it. In addition, some of the outflow was the repatriation by foreigners of investments in the U.S.[23] Nevertheless, U.S. overseas investment must have also grown rapidly in these years

**Table 2.3: U.S. Exports, Direct Investments, and Investment Income, 1900–13**

| *Year* | *Exports*[a] | *Investments*[b] ($ Millions) | *Income*[c] | *Exports* | *Investments* (Index, 1900=100) | *Income* |
|---|---|---|---|---|---|---|
| 1900 | 1,686 | 56 | 38 | 100 | 100 | 100 |
| 1901 | 1,651 | 89 | 47 | 98 | 159 | 124 |
| 1902 | 1,550 | 65 | 57 | 92 | 116 | 150 |
| 1903 | 1,663 | 81 | 67 | 99 | 145 | 176 |
| 1904 | 1,657 | 80 | 70 | 98 | 143 | 184 |
| 1905 | 1,859 | 46 | 76 | 110 | 82 | 200 |
| 1906 | 2,052 | 92 | 86 | 122 | 164 | 226 |
| 1907 | 2,192 | 89 | 87 | 130 | 159 | 229 |
| 1908 | 2,022 | 48 | 89 | 120 | 86 | 234 |
| 1909 | 2,013 | 88 | 100 | 119 | 157 | 263 |
| 1910 | 2,160 | 124 | 108 | 128 | 221 | 284 |
| 1911 | 2,405 | 95 | 114 | 143 | 170 | 300 |
| 1912 | 2,738 | 139 | 123 | 162 | 248 | 324 |
| 1913 | 2,816 | 138 | 137 | 167 | 246 | 361 |

a. Exports of goods and services.
b. Private investments only.
c. Includes income from both direct and *rentier* investments.

Sources: *Historical Statistics of the United States, Colonial Times to 1957: A Statistical Abstract Supplement* (U.S. Government Printing Office, Washington, 1960), pp. 562, 564.

because, as Table 2.3 indicates, the U.S. accounts recorded an investment income of $38 million for 1900. Presumably most of that income came from investment in previous years. Furthermore the official U.S. accounts recorded, for the first time, data on U.S. capital outflows for 1900,[24] indicating that they had reached such a high level by that year that they warranted entry as a separate item.

*Direct Investment*

Most of this outflow was direct, as opposed to *rentier*, investment. Direct investment gives the investor control over resource use while *rentier* normally does not. Direct investors are mainly large corporations who seek to control production facilities in other countries. *Rentier* investors are normally wealthy individuals or large financial institutions who seek to maximize income from a portfolio. They are usually more interested in shifting assets within their portfolios in order to obtain a maximum yield than in acquiring control over a particular company, since the acquisition of such control may hinder their flexibility in shifting investments from one company to another.

As Table 2.3 shows, U.S. direct investments grew at a much faster rate than exports between 1900 and 1913, the last full year of peace before the outbreak of World War I. Income from investments, including *rentier* as well as direct, rose at a much steeper pace than either exports or investments. Presumably most of this increased income came from the direct, as opposed to the *rentier*, investments.

The reason for this presumption is that direct investments can occur only in a world of imperfect competition, *i.e.*, in a world dominated by large firms which usually sell differentiated products. As Kindleberger notes:

> . . . in a world of perfect competition of goods and factors, direct investment cannot exist. In these conditions, domestic firms would have an advantage over foreign firms in the proximity of their operations to their decision making centers, so that no firm could survive in foreign operation. For direct investment to thrive there must be some imperfection in markets for goods or factors, including among the latter technology . . .[25]

Direct investment enables large firms to exploit these market imperfections by earning a higher average rate-of-return.

As was pointed out earlier in this chapter, the domestic expansion of the large U.S. firms was characterized by the acquisition of production facilities in diverse geographic areas. This direct investment at the domestic level became the international model by the turn of the century.

**Table 2.4: Distribution of U.S. Direct Private Foreign Investments by Industry, 1914**

| *Industry* | *Investment* ($ Millions) | *Per Cent* |
|---|---|---|
| Mining and Smelting | 720 | 27 |
| Manufacturing | 478 | 18 |
| Agriculture[a] | 356 | 13 |
| Petroleum | 343 | 13 |
| (Distribution) | (200) | ( 8) |
| (Production[b]) | (143) | ( 5) |
| Railroads | 255 | 10 |
| Other[c] | 500 | 19 |
| Total | 2,652 | 100 |

a. Includes investments in timber.
b. Includes refining.
c. Sales organizations, utilities, banks, and insurance companies.

Sources: C. Lewis, *America's Stake in International Investment* (Brookings Institution, Washington, 1938), p. 578 ff. and M. Wilkins, *The Emergence of Multinational Enterprise: American Business Abroad from the Colonial Era to 1914* (Harvard University Press, Cambridge, Mass., 1970), p. 110.

In 1897, for instance, the book value of U.S. private direct foreign investment was estimated at $600 miliion. By 1908 this value had risen almost threefold to $1.6 billion (thousand million). By June 30, 1914, on the eve of World War I, the estimated book value had increased to $2.7 billion,[26] more than a fourfold increase over the 1897 level. As can be seen in Table 2.4, at the outbreak of World War I the bulk of this investment, over two-thirds, was in four industries: mining, manufacturing, agriculture, and petroleum. Table 2.5 indicates that almost three-quarters of this investment was concentrated in the Western Hemisphere, an area over which the U.S. exercised hegemony. Europe, the only other area receiving substantial investments, was, at that time, outside the area of U.S. control.

Most of the investment in Europe was in manufacturing, chiefly in those industries in which U.S. firms had either a technological advantage or a differentiated product. Among the firms with manufacturing plants in Europe in 1914 were American Radiator, American Tobacco, Diamond Match, Ford, General Electric, Heinz, International Harvester, National Cash Register, Singer Sewing Machine, Western Electric, and Westinghouse.[27] All these firms had either a specific

**Table 2.5: Geographical Distribution of U.S. Direct Private Foreign Investment, 1914**

| *Region* | *Investment* ($ Millions) | *Per Cent* |
|---|---|---|
| Western Hemisphere | 1,899 | 72 |
| (Canada) | (618) | (23) |
| (Mexico) | (587) | (22) |
| (Caribbean and Central America) | (371) | (14) |
| (South America) | (323) | (12) |
| Eastern Hemisphere | 753 | 28 |
| (Europe) | (573) | (22) |
| (Asia[a]) | (120) | ( 5) |
| (Other[b]) | ( 60) | ( 2) |
| Total[c] | 2,652 | 100 |

a. Excludes investments by Standard Oil (New Jersey) in the Dutch East Indies.
b. Africa, Oceania, and "unassigned."
c. Because of rounding the total may not equal the sum of the column.

Sources: C. Lewis, *America's Stake in International Investment* (Brookings Institution, Washington, 1938), p. 578 ff. and M. Wilkins, *The Emergence of Multinational Enterprise: American Business Abroad from the Colonial Era to 1914* (Harvard University Press, Cambridge, Mass., 1970), p. 110.

technological advantage or a unique product from which they hoped to reap a profit in the European market. They chose the path of direct investment in Europe, rather than exporting from the U.S., for a number of reasons: high product transportation costs, tariffs in European countries, differences in wage rates, and the buttressing of their competitive positions *vis-à-vis* their actual and potential European rivals. Furthermore, advances in technology in communications and transport, like the laying of the trans-Atlantic cable and speedier steamship crossings of the ocean, enabled the top decision-makers of a firm to exercise effective control of operations in geographically dispersed areas, giving a further impetus to the drive to acquire overseas production facilities.

The other area of investment in manufacturing was Canada. Manufacturing investment there in 1914 was greater than in Europe. In that year 46 per cent of manufacturing investment was in Canada as opposed to 42 per cent in Europe.[28] Manufacturing investment flowed to Canada essentially for the same reasons that it flowed to Europe. But, unlike Europe, Canada was also the recipient of investment in the primary industries. As Table 2.6 shows, 46 per cent of U.S. investment in Canada was in mining, agriculture, and petroleum production while only 2 per cent was in those industries in Europe.

U.S. investment in Canada was, however, relatively small when compared with British investment. As Table 2.7 indicates, English interests in 1913 owned almost three-quarters of all foreign investments in the country while U.S. interests held less than one-quarter. Over 80 per cent of the U.K. investment was portfolio, or *rentier*, investment. U.S. investments, in contrast, were direct. Thus, on the eve of World War I, foreign capital in Canada consisted mainly of debt capital controlled by British interests. As we shall see later, this pattern was to change dramatically in subsequent years.

In the rest of the Western Hemisphere U.S. investment was largely

**Table 2.6: Distribution of U.S. Direct Private Foreign Investment by Region and by Industry, 1914**

| *Industry* | *Europe* (%) | *Canada* (%) | *Latin America*[a] (%) |
|---|---|---|---|
| Manufacturing | 35 | 36 | 3 |
| Petroleum | 24 | 4 | 10 |
| (Distribution) | (23) | – | ( 2) |
| (Production) | ( 1) | ( 4) | ( 8) |
| Sales Organizations[b] | 15 | 4 | 3 |
| Utilities | 2 | 1 | 8 |
| Mining and Smelting | 1 | 26 | 43 |
| Railroads | – | 11 | 14 |
| Agriculture[c] | – | 16 | 19 |
| Other[d] | 23 | 1 | 1 |
| Total[e] | 100 | 100 | 100 |

a. Includes Mexico, Central America, the Caribbean countries, and South America.

b. Excludes petroleum distribution. Includes trading companies and sales branches and subsidiaries of large corporations.

c. Includes timber.

d. Mainly banks and insurance companies.

e. Because of rounding the total may not equal the sum of the column.

Sources: Based on C. Lewis, *America's Stake in International Investment* (Brookings Institution, Washington, 1938), pp. 579 and 588, and M. Wilkins, *The Emergence of Multinational Enterprise: American Business Abroad from the Colonial Era to 1914* (Harvard University Press, Cambridge, Mass., 1970), p. 110.

**Table 2.7: Foreign Capital Invested in Canada, 1913**

| *Capital Source* | *Total* ($ Millions[a]) | (%) | *Portfolio* ($ Millions[a]) | (%) | *Direct* ($ Millions[a]) | (%) |
|---|---|---|---|---|---|---|
| United Kingdom | 2,818 | 73 | 2,618 | 85 | 200 | 26 |
| (Per cent) | (100) | | (93) | | (7) | |
| United States | 835 | 22 | 315 | 10 | 520 | 68 |
| (Per cent) | (100) | | (38) | | (62) | |
| Other | 197 | 5 | 147 | 5 | 50 | 6 |
| (Per cent) | (100) | | (75) | | (25) | |
| Total | 3,850 | 100 | 3,080 | 100 | 770 | 100 |
| (Per cent) | (100) | | (80) | | (20) | |

a. Canadian dollars.

Source: Calculated from K. Levitt, *Silent Surrender, The Multinational Corporation in Canada* (Macmillan of Canada, Toronto, 1970), p. 67.

confined to the primary industries. Investment in Latin America was concentrated in three primary industries (mining, agriculture, and petroleum) and railroads. These four industries accounted for over 85 per cent of U.S. investment in that area. It is interesting to note the difference in investment patterns in Latin America, on the one hand, and Europe and, to a lesser extent, Canada, on the other. In that area over which the U.S. exercised hegemony, Latin America, the investments were essentially in the extractive industries. Even the investments in railroads, which were concentrated in Mexico, were related to the extraction of minerals from that country. Contrary to the pattern in Europe and Canada, very little investment went into manufacturing.

The U.S. investment pattern was a typically colonial one. This pattern consisted of assigning manufacturing industries to the "mother" country and agricultural and raw-material producing industries to the conquered areas. Such was the essence of the mercantilist policies adopted by the major European powers in the sixteenth and seventeenth centuries. In a similar manner, the U.S. confined its investments in those areas over which it exercised political control, like Latin America, to agriculture and raw materials. In those areas over which the U.S. had no political control, like Europe, or only partial political control, like Canada, a substantial portion of its investments went into manufacturing.

Thus the U.S. investors looked to Latin America as the supply source for their manufacturing base, and to Europe and to Canada as outlets for their investments in manufacturing. Their manufacturing investments in Canada, however, were confined to the already developed regions of that country. In the rest of the country their investments followed the colonial pattern.

By 1914 U.S. firms were playing an increasingly important role in the world economy as a whole and in the Western Hemisphere in particular. The bulk of the U.S. investments were in the extractive industries and in industries directly linked to them, like railroads. Among the extractive industries traditional mining, for example of nonferrous metals, was the most important. By 1914, however, a new industry, oil, had come onto the scene. This industry was destined to play an increasingly important role in the future external expansion of the United States.

### Notes

1. *Historical Statistics of the United States, Colonial Times to 1957: A Statistical Abstract Supplement* (U.S. Government Printing Office, Washington, 1960), p. 236.
2. The full extent to which the United States was involved in wars against the Indians was revealed in a 1950 publication of the U.S. Armed Forces Information School. This publication lists ninety-nine "wars, expeditions, campaigns, occupations, and other disturbances in which the Army of the United States has participated" in the period between 1775 and 1950. This count does not include the Korean or Vietnam wars nor separate military activities conducted by the Navy and the Marines. The bulk of these ninety-nine "wars, etc." took place in the nineteenth century against the Indians. They included such long wars as "The Apache Indian Wars and Troubles," which lasted twenty-nine years; "The Navajo Troubles" and the "Continuous Disturbances with Comanche, Cheyenne, Lipan and Kickapo Indians," each of which lasted twelve years; and the four "Seminole Wars," which spanned a period of over eighty years. See Armed Forces Information School, *The Army Almanac* (U.S. Government Printing Office, Washington, 1950), pp. 439–41.
3. *Historical Statistics of the United States*, p. 236.
4. The liquidation of the Indian population was apparently an unwritten objective of the various U.S. governments during this period. This objective was translated into the folklore through the popular expression, "The only good Injun is a dead Injun." There are no accurate figures on the effectiveness of this genocidal policy. Some estimates place the Indian population of what is now the Continental United States at 1,000,000 in 1800. The first

complete census of Indians, taken in 1890, placed their number at around 250,000. See *Historical Statistics of the United States*, p. 9.

5. A trust was a type of combination in which stockholders in competing firms surrendered their stock to a group of trustees in exchange for trust certificates. The trustees were then able to control the competing firms and, in effect, to operate them as one unit. The higher profits resulting from these monopolistic operations were distributed to the owners of the trust certificates as dividends.

   The trust was devised by one of Rockefeller's lawyers. Following Rockefeller's initiative in oil, trusts were formed in such industries as whisky, sugar, lead, linseed oil, and several others. The unpopularity of the trusts made them the object of hostile legislation and court decisions. As a result, the trusts were abandoned in favour of the holding company.

   The holding company is a device by which one company controls another company through the ownership of stock. Often a holding company can gain control over another company by purchasing only a small share of the latter company's stock. Under legislation passed in the early 1900s, trusts were enabled to become holding companies by transforming the trust certificates into stocks. It was under this legislation that the Standard Oil Trust was transformed into the holding company, Standard Oil of New Jersey.

   Each company controlled by a holding company retains its identity. With a merger one company normally buys out the stocks or assets of the other. The bought company usually loses its identity; it is absorbed by the buyer.

6. F. M. Scherer, *Industrial Market Structure and Economic Performance*, (Rand McNally, Chicago, 1970), p. 48.
7. S. Ryerson, *Unequal Union: Roots of the Crisis in the Canadas, 1815–1873* (Progress Books, Toronto, 1973), p. 340.
8. Ibid.
9. Ibid.
10. Ibid.
11. Ibid., p. 341.
12. Ibid.
13. S. Nearing & J. Freeman, *Dollar Diplomacy*, reprint (Monthly Review Press, New York and London, 1966), p. 176.
14. Ibid., p. 177.
15. The U.S. Constitution bars from the presidency anyone born outside the United States.
16. W. A. Williams, *The Roots of the Modern American Empire: A Study of the Growth and Shaping of Social Consciousness in a Marketplace Society* (Random House, New York, 1969), p. 365.
17. Ibid., p. 365.
18. Ibid., pp. 440–41.
19. These landings and occupations are too numerous to document in this book. They are described in such works as Nearing & Freeman, *Dollar Diplomacy*; Williams, *The Roots of the Modern American Empire*; and M. Wilkins, *The Emergence of the Multinational Enterprise* (Harvard University Press, Cambridge, Mass., 1970).

20. "President Taft Defines Dollar Diplomacy, December, 1912" in W. A. Williams (ed.), *The Shaping of American Diplomacy: Readings and Documents in American Foreign Relations* (2 vols., Rand McNally, Chicago, 1956), vol. 2, pp. 494–95.
21. C. K. Hobson, *Export of Capital* (Constable, London, 1914), chaps. 1 and 2, and V. I. Lenin, *Imperialism: The Highest Stage of Capitalism* (International, New York, 1969), p. 62.
22. *Historical Statistics of the United States*, p. 564.
23. Ibid., p. 558.
24. Ibid., p. 562.
25. C. P. Kindleberger, *American Business Abroad: Six Lectures on Direct Investment* (Yale University Press, New Haven and London, 1969), p. 13.
26. *Historical Statistics of the United States*, p. 565.
27. M. Wilkins, *The Emergence of Multinational Enterprise: American Business Abroad from the Colonial Era to 1914* (Harvard University Press, Cambridge, Mass., 1970), pp. 212–13.
28. Ibid., p. 110.

CHAPTER THREE

# The Oil Industry's Role in American Expansion

The U.S. oil industry came onto the scene in 1859, when "Colonel" Edwin Drake drilled the nation's first oil well near Titusville, Pa.[1] Though oil was sold commercially in Canada, Russia, Burma, and Rumania before 1859,[2] Drake's well was the one that was destined to have the greatest impact on the world. The success of that well was due to a number of factors.

First, Drake, like the typical Schumpeterian innovator, introduced a new technique of production. He drilled for oil instead of digging for it.[3] Drilling reduced significantly the cost of extracting oil, thus making oil competitive with other fuels. Second, technological advances provided a ready market for this oil. Lamps designed to burn oil distilled from coal were a recent such advance. These lamps were widely distributed throughout the U.S., providing a ready outlet and marketing network for the kerosine derived from crude oil. Furthermore, the coal-oil industry, which was the fastest growing illumination industry in the U.S. during the 1850s,[4] had developed techniques for refining, which the oil industry was able to adopt.[5] Third, the U.S. was growing at a rapid pace. The outbreak of the Civil War in 1861 greatly accelerated this pace. The expanding industrial base required ever greater quantities of illuminants and lubricating oils, both of which the fledgling oil industry could provide at relatively low prices. Furthermore, this industrialization was not confined merely to the United States. It was also taking place in Western Europe, which provided an important market for the U.S. oil industry. The U.S. was able to corner this market because its fields were far more prolific than those of its closest rival, Canada. In 1862, for instance, the U.S. fields produced more than three million barrels while the Canadian ones produced only seventy thousand.[6] In fact, in that year all the world's oil was produced in the United States and Canada. The U.S. share was 97 per cent, the Canadian, 3 per cent.[7] Thus this conjuncture of lower costs, new technology, growing industrialization, and prolific fields had the effect of making Drake's well the one that ushered in the oil era.

## The Organization of the Industry

With the growth of this new industry, there also emerged a new group of entrepreneurs, who were eager to control it and propel it forward. These entrepreneurs engaged in both interproduct and intraproduct rivalry: they sought markets not only in areas held by suppliers of other types of fuel, like coal-oil, but in areas held by each other. The very logic of this intense rivalry, as Schumpeter has so aptly pointed out,[8] is for each entrepreneur to strive to obtain monopoly positions, which will both protect him against attacks from others and enable him to attack his rivals more effectively.

This process of intense competition followed by monopolization became evident in the early years of the oil industry. This process, it should be noted, was independent of the will of the actors. It can be presumed that most, if not all, actors desired to achieve a monopoly position. However, the ability of any one actor to do so was dependent not only on his individual business acumen but also on the objective conditions in that particular industry.

### *Objective Conditions*

In the early development of the oil industry, monopolization was made possible by the presence of three conditions. They were (1) the presence of scale economies in refining, (2) the geographical distance between the major markets and the producing centres, and (3) the limited supplies of crude oil available.

#### SCALE ECONOMIES

Scale economies occur when average production costs fall as the size of the production unit increases. The presence of significant scale economies in any industry is incompatible with perfect competition in that industry. If the optimum plant size (the plant size at which average costs are at a minimum) is large in relation to the size of the market, then that market can support at most only a few plants or, in the extreme case of "natural monopoly," only one. Scale economies spring from a number of sources, including technology, greater specialization, and a higher utilization of scarce factors like skilled management.

Scale economies appeared in the U.S. refining industry at a relatively early date. Though the data are scanty, the evidence nevertheless indicates that average costs were lower in the larger refineries. In the 1880s, for instance, the average refinery had a capacity of between 1,500 and 2,000 barrels per day. Its average costs was about 2.5 cents per gallon. The costs in the refineries owned by the Standard Oil Trust were stated by the company to be 0.534 cents per gallon.[9] The Trust's refineries were much larger than those of its competitors. The largest one, located at Bayonne, N.J., had a capacity of between 6,000 to 7,000 barrels per day.[10] Some critics claim that Standard Oil deliberately understated its cost figures to give the impression that its dominant position was due solely to its superior efficiency; further, these alleged efficiencies may have been in part due to factors other than scale economies. There are nevertheless indications that scale economies were important. One such indication was the declining number of refineries. Though the demand for products was rising, the number of refineries operating in the U.S. fell from 155 in 1870 to 89 in 1880.[11] This absolute reduction in the face of rising demand could not have occurred unless large-scale units reduced the possible number of competitors, thereby creating conditions for the monopolistic control of the industry.

### GEOGRAPHICAL DISTANCE

When markets are geographically isolated from sources of supply, transportation costs may become an important factor in determining the specific structure of a given industry. To the extent that transport costs are a significant component of the final price, then a reduction in these costs can give one firm a marked advantage over its competitors. The ability of a firm to achieve advantageous freight rates rests both on its own bargaining position and on the structure of the transportation industry. If perfect competition exists in a transportation industry, it is impossible for a firm to give a particular shipper more favourable terms than another. Competition will eventually force all firms to charge all shippers the same price. If, on the other hand, competition does not exist, then a firm can discriminate among shippers. The firm in fact has an incentive to favour larger shippers over smaller ones; in return for a guarantee of a constant, high volume of shipments, it will lower rates. Such bilateral arrangements offer many advantages to both parties.

The geographical separation of markets from sources characterized the early U.S. oil industry. The main sources were in western Pennsylvania and Ohio. The principal markets for refined products were in New England, the Middle Atlantic States, and Europe.[12] Furthermore, considerable refining capacity was concentrated in areas like Pittsburgh and Cleveland, which were located between the oil fields and the markets. In 1873, these two cities alone accounted for almost one-half of U.S. refining capacity.[13] Oil going to and from these refining centres thus incurred two transportation charges, the charge on the crude oil flowing from the fields to the refineries and the charge on the products going from the refineries to their ultimate destinations. These charges were high relative to the value of the goods. For instance, in the early 1870s, the Standard Oil Trust was paying approximately twenty per cent of value for both its crude and product shipments.[14] And Standard received favourable rates; its competitors had to pay considerably more for transportation.

The chief means of transporting crude were the railroads and pipelines. They were also the chief means of transporting products. In addition, the Cleveland refineries were able to transport products by boat to New York via the Great Lakes and the Erie Canal. Since none of these industries was perfectly competitive, an environment existed which nurtured bilateral arrangements between large shippers of oil and the transportation companies. These arrangements played a major role in shaping the future structure of the oil industry.

### LIMITED SUPPLIES

The existence of a finite stock of a raw material input creates both the possibility of and the incentive for monopoly control. A monopolist controlling that stock has no fear of new entrants. Furthermore, he can exercise control over the industries using the raw material as an input. Control of this finite stock therefore can give him considerable leverage throughout the economy. The very existence of such a stock thus engenders powerful incentives to control it.

This control can be achieved in two ways. One is through contracts and other arrangements for the purchase of raw materials. The other is through buying out the owners of the stock. In the early history of the U.S. oil industry, the former method was used. In later periods, the oil companies attempted to use the latter.

In the years between the drilling of Drake's well and the turn of the century, practically all the oil in the U.S. was located in the fields of western Pennsylvania and Ohio. Though the ownership of wells in these fields was dispersed among fourteen thousand producers,[15] the control of production was actually in the hands of a few large refiners, who bought the bulk of crude production. These refiners had little to fear from competition as long as the bulk of the crude came from these fields. These limited supplies satisfied one of the key requirements for monopoly control.

*Business Acumen*

Objective conditions are a necessary but not sufficient condition to bring about monopoly control. What is also needed are entrepreneurs who can recognize these conditions and utilize them for their own ends. Presumably the existence of these conditions will, in and of itself, give birth to these entrepreneurs.

Such an entrepreneur was John D. Rockefeller. The role of the entrepreneur, according to Schumpeter, is to create "the new consumers' goods, the new methods of production or transportation, the new markets, the new forms of industrial organization."[16] Because Rockefeller had a better understanding than anyone else of the objective conditions in the oil industry, in a relatively short time he was able to create a "new form of industrial organization" by transforming the industry from a competitive one to a monopoly. In doing so he used methods which earned him universal condemnation. Though his severest critics attributed his successes to his questionable practices, they failed to realize that his methods were thoroughly consistent with the mores of competitive capitalism and, further, that his methods could not have been successful had not the objective conditions for their success been present. If Rockefeller had not changed the organization of the industry, someone else would have.

A great deal has been written about Rockefeller's tactics. They include such practices as manipulation of freight rates, industrial espionage, and threats against distributors who bought from his competitors. Through both promises to guarantee the railroads a minimum quantity of shipments and threats to build pipelines if the railroads did not comply, he obtained rebates on his shipments and drawbacks[17] on the shipments of his competitors.

Although Rockefeller tried to keep his agreements secret, he was often unable to do so, and many were leaked to the public. One such agreement was the South Improvement Company Plan of 1872. Though the Pennsylvania State Legislature declared that plan, with its highly discriminatory freight rate structure, invalid shortly after its adoption, Rockefeller nevertheless was able to use it to obtain control of the Cleveland refining industry. Ida Tarbell, one of his early biographers, describes how he did so:

There were at that time some twenty-six refineries in the town—some of them very large plants. All of them were feeling more or less the discouraging effects of the last three or four years of railroad discriminations in favour of the Standard Oil Company. To the owners of these refineries Mr. Rockefeller now went one by one, and explained the South Improvement Company. "You see," he told them, "this scheme is bound to work. It means an absolute control by us of the oil business. There is no chance for anyone outside. But we are going to give everybody a chance to come in. You are to turn over your refinery to appraisers, and I will give you Standard Oil Company stock or cash, as you prefer, for the value we put upon it. I advise you to take the stock. It will be for your good." Certain refiners objected. They did not want to sell. They did want to keep and manage their business. Mr. Rockefeller was regretful, but firm. It was useless to resist, he told the hesitating; they would certainly be crushed if they did not accept his offer. . .[18]

Within three months, twenty-one of the twenty-six refineries sold out to Rockefeller. Standard Oil's capacity rose from approximately 1,500 barrels per day to 10,000 barrels. By 1872, Rockefeller's Standard Oil Company, which had been incorporated only two years earlier, owned more than one-fifth of the United States' refining capacity.[19] The ownership of numerous refineries gave Rockefeller the power to rationalize production by shutting down the small, inefficient units and concentrating it in the larger, more efficient ones. Between 1882 and 1886, for instance, he closed down thirty-one refineries, confining production to his twenty-two larger units.[20] This increased efficiency gave Standard an added advantage over its competitors, enabling it to garner a larger share of the market.

The very size of Standard Oil gave it additional power to win further concessions from the railroads. This, in turn, further increased its size and power. By 1899, it owned 82 per cent of U.S. refining capacity.[21] This extensive ownership gave it effective control over crude supplies.

In the years immediately following the discovery of oil in Pennsylvania, crude producers used to sell their wares to refiners through the Oil City Exchange. This Exchange, functioning like the ideal markets described in textbooks, enabled buyers and sellers to come together. As a result of their transactions a market price was established, which often fluctuated wildly. Perhaps more important than its price role, the Exchange enabled independent crude producers to find markets and independent refiners to find sources of supply.

This particular function became less relevant as the Exchange began to be dominated by one larger buyer, Standard Oil. The size of Standard forced most independent crude producers to sell to it. Nevertheless, some individual producers were able to establish contacts with independent refiners through the Exchange. This opportunity for establishing contacts disappeared in 1895 when Standard decided to bypass the Exchange and buy directly from producers in the field at a price posted by it. This action forced the Exchange to close,[22] thereby making it extremely difficult for independent refiners to contact producers.

Another method Rockefeller used to foreclose oil from independent refiners was to enter into production-sharing arrangements. In the late 1880s, when crude prices were falling because of increased production, he agreed to stabilize prices provided producers reduced their output. Prices rose as wells were shut in.[23] The shutting down of wells decreased the quantities flowing to independent refiners. At the same time, the rise in prices encouraged exploration in other Pennsylvania fields, which Rockefeller sought to control. This control of new oil was necessary both for replacing oil from the older, depleted wells and for keeping it out of the reach of independent refiners. While Rockefeller never did achieve complete control over the Pennsylvania and Ohio fields and while he was unable to prevent the rise of a competitor, Pure Oil, he nevertheless managed to gain control of 88 per cent of Pennsylvania and 85 per cent of Ohio crude production by the turn of the century.[24] Because of this, he was able to restrict the expansion of rivals like Pure Oil.

The size of Standard also made most distributors of petroleum products vulnerable to threats by the company. Rockefeller used this power to stop them from buying from his competitors. To obtain information on distributors, he established an elaborate network of industrial espionage. Included in this network were employees of the railroads, who furnished him with data on shipments by his competitors. Ida Tarbell, who in 1903 came into possession of reports filed by Standard's agents, described how the network operated:

> They [the documents] show that the Standard Oil Company receives regularly today, at least from the railroads and steamship lines represented in these papers, information on *all* oil shipped. A study of these papers shows beyond question that somebody having access to the books of the freight offices, records regularly each oil shipment passing the office—the names of consignor and consignee, the addresses of each, and the quantity and kind of oil are given in each case. . . . Now what is this for? Copies of letters and telegrams accompanying the reports show that as soon as a particular report had reached Standard headquarters and it was known that a carload, or even a barrel, of independent oil was on its way to a dealer, the Standard agent whose name was written after the shipment on the record had been notified.

"If you can stop car going to X, authorise rebate to Z (name of dealer) of three-quarters cent per gallon," one of the telegrams reads. There is plenty of evidence to show how an agent receiving such information "stops" the oil. He *persuades* the dealer to countermand the order.[25]

Both the carrot and the stick were used as methods of "persuasion." Initially the dealer was offered concessions if he countermanded the order. If he refused, he was threatened with price wars. Tarbell cites letters submitted to the State of Ohio's 1898 inquiry into the practices of Standard Oil. A typical letter from a dealer to an independent distributor or refiner read:

> I am sorry to say that a Standard Oil man from your city followed that oil car and oil to my place, and told me that he would not let me make a dollar on that oil, and was dogging me around for two days to buy that oil, and made all kinds of threats and talked to my people of the house while I was out, and persuaded me to sell, and I was in a stew what I should do, but I yielded and I have been sorry for it since. I thought I would hate to see the bottom knocked out of the prices, but that is why I did it—the only reason.[26]

Rockefeller's tactics enraged not only the "little" man but some members of the "elite" as well. The Wall Street banker John P. Morgan made no secret of his dislike of Rockefeller.[27] This dislike probably stemmed more from his business dealings with Rockefeller than from his disapproval of Rockefeller's tactics. Morgan's own tactics did not, after all, differ significantly from those of Rockefeller. They were the tactics of the rising American elite, these Schumpeterian entrepreneurs, who were out to dominate and control.

The elite, which had considerable control over the press, constantly attacked Rockefeller through such papers as Hearst's *New York Journal* and Pulitzer's *New York World*.[28] They resented this newcomer to their ranks, who was threatening their dominant position in society. They were joined by such populists as U.S. Senator Robert LaFollette of Wisconsin, who called Rockefeller "the greatest criminal of the age."[29]

It was this conjuncture of forces that led the U.S. government to launch its famous anti-trust action against Standard Oil. The case finally ended in 1911, when the Supreme Court upheld the decision of the lower court ordering the dissolution of the company. In this decision, which actually served to strengthen monopoly power in the U.S., the Court found that

> the very genius for commercial development and organization which it would seem was manifested from the beginning soon begot an intent and purpose to exclude others which was frequently manifested by acts and

dealings wholly incompatible with the theory that they were made with the single conception of advancing the development of business power by usual methods, but which, on the contrary, necessarily involved the intent to drive others from the field and to exclude them from the right to trade, and thus accomplish the mastery which was their end in view.... The exercise of the power which resulted from that organization fortifies the foregoing conclusions, since the developments which came, the acquisition here and there which ensued of every efficient means by which competition could have been asserted, the slow but relentless methods which followed by which means of transportation were absorbed and brought under control, the system of marketing which was adopted by which the country was divided into districts and the trade in each district in oil was turned over to a designated corporation within the combination, and all others excluded, all lead the mind up to a conviction of a purpose and intent which we think is so certain as practically to cause the subject not to be within the domain of reasonable contention.[30]

The Court stressed the term "reasonable contention" because it found Standard Oil's actions in restraint of trade to be "unreasonable." Presumably, if the actions were "reasonable" the Court would not have ruled against the company. By adopting this "rule of reason" the Court gave itself wide powers to determine whether or not a company should be penalized for its behaviour. These powers enabled the Court to take sides in the various disputes among factions of the American elite. Thus, the Court in 1920 used the same "rule of reason" to absolve the United States Steel Corporation of violations of the Anti-Trust Law, even though the Court admitted that the corporation engaged in practices which "may be violations of the law."[31] The Court found these practices to be "reasonable" because they were "transient in their purpose and effect."[32] The United States Steel Corporation, it should be noted, was established by J. P. Morgan and controlled by Morgan interests. The differential treatment of Rockefeller and Morgan probably was a reflection of power relationships within the American establishment. The Rockefellers at that time had still not reached the apex of power.

Though the Court dissolved the Standard Oil Company of New Jersey, it did not make any significant dent in Rockefeller's armour. The decision served more to slow his rise than to topple him. The judgement severed Standard's relationship with its subsidiaries by ordering Standard to deliver directly to its stockholders the shares it held in the subsidiaries.[33] In effect this meant that Rockefeller, who held 244,385 shares of Standard's 983,383 shares,[34] received proportionate shares in the thirty-nine constituent companies. For many years after dissolution, the former Standard subsidiaries worked in harmony with each other, though eventually some did strike out on their own. While the decision did stunt Standard's growth, it did not significantly destroy the harmony of interests of the Standard companies, which Rockefeller had so laboriously built up.

In reality, even before the decision, certain events were unfolding, which served to temper the hostility towards Rockefeller. These events flowed from changes in competitive conditions at home and the rise of new companies abroad.

*Changes in Domestic Competitive Conditions*

Despite all his efforts, Rockefeller was not able to exclude newcomers from the industry. His strategy had been to own the refineries and the transportation facilities, leaving the production of crude to others. This strategy was based on the premise that there was only a limited amount of crude, whose production he could control through the ownership of downstream facilities, like refining and distribution. For the first fifteen years of Standard's corporate life, therefore, he neglected opportunities to invest in new sources of crude. In 1888, for instance, Standard produced from its own properties less than 200 barrels per day of crude. Total production in the U.S. was around 76,000 barrels per day.[35]

Rockefeller also felt that production of crude was basically unprofitable. He changed his mind when crude prices began to rise in the late 1880s, and he began acquiring producing properties. By 1891 Standard produced more than one-quarter of domestic crude.[36] Nevertheless it did not acquire crude as vigorously as did other facilities. In the Supreme Court case Standard's attorneys argued that the company was not a monopoly because it controlled "a very small percentage of the crude oil produced."[37] Probably the major reason for this "very small percentage" was Standard's failure to acquire producing properties in Texas.

The U.S. oil picture changed dramatically as a result of the Spindletop discovery in Texas in 1901, one of the great discoveries of all time. Not only did it significantly increase the supply of oil but it also provided the opportunity for new companies to enter.

These new entries may never have occurred had Standard ignored the advice of its "expert." In 1900 Standard was approached by Captain Anthony Lucas,[38] who was drilling for oil at Spindletop, for financial backing. The company sent Calvin Payne, its authority on oil properties, to survey the terrain. In what was probably one of the most colossal misjudgements in history he reported that there was "no indication whatever to warrant the expectation of an oil field on the prairies of southeastern Texas" and that Spindletop "has no analogy to any oil field known, as in fact there was not the slightest trace of even an oil escape."[39]

There was another reason why Rockefeller refused to invest in Spindletop. The State of Texas, spurred by one of its leading politicians, James Hogg, had launched legal actions against the Rockefeller companies operating within its boundaries. It even passed a law which effectively prohibited Rockefeller from doing any business in the state.[40] Rockefeller was determined to "punish" Texas by refusing to invest in it. Thus, even after Spindletop was proven to be a success, Standard turned down a chance to buy into it. Representatives of the company reportedly told the Mellon family of Pittsburgh, who approached Standard to sell to them the Mellon investment in Spindletop, that "after the way Mr. Rockefeller has been treated by the state of Texas, he'll never put another dime in Texas."[41]

As it turned out, it was Rockefeller, rather than Texas, who was "punished" by this decision. The Mellons, who financed Spindletop after Rockefeller initially refused to do so, stayed in the oil business and organized Gulf Oil, which became a major competitor of Standard. In addition, Texas financial interests brought together by Jim Hogg established the Texas Company, which also threatened Standard's supremacy. The possession by both of these companies of large reserves of Texas oil gave them a distinct advantage over Standard, with its relatively small crude supplies.

By the time of its dissolution in 1911, Standard controlled only 10 per cent of oil in the fields of the Texas Gulf coast. Its control of refinery capacity had fallen to 64 per cent from a high of approximately 95 per cent thirty years earlier.[42] The new entrants, whose power was based on the ownership of crude rather than refining capacity, were able to breach Rockefeller's domestic monopoly even before the Supreme Court acted. Furthermore, new entrants abroad, whose power was also based on the possession of crude, posed an additional threat to Rockefeller's dominance of foreign markets.

### *New Companies Abroad*

In the latter part of the nineteenth century, two important new oil centres arose in Russia and in Indonesia. Russian oil was controlled by the Nobel and the French Rothschild interests, while Indonesian oil was controlled by Dutch and English interests. These groups competed vigorously among themselves and with Rockefeller for control over the world's supplies and markets. In this struggle, each of these groups was backed by its home government. Thus, the U.S. government, while ostensibly punishing Rockefeller for his actions at home, gave considerable support to his activities abroad. This support was an essential ingredient in its policy of expanding exports.

## Foreign Trade

From its very beginning oil was important in the United States' foreign trade. In 1866, only seven years after the Drake well and only three years after John D. Rockefeller had entered the oil business, John sent his brother William to New York to set up an export office.[43] At this time exports of refined products already exceeded domestic sales. In 1866, more than two-thirds of refinery output was exported. By 1873 exports accounted for almost three-quarters of refinery output.[44] This high proportion of exports continued almost to the turn of the century, when two developments—the rise of competition abroad and the growth of the domestic market—reduced it.

Another aspect of the foreign trade in oil is that it was mainly trade in refined products rather than in crude. This pattern was strikingly different from that exhibited in the rest of America's foreign trade. By the end of the U.S. Civil War, American industrialization lagged behind European. The U.S. continued to play its historical role of an exporter of raw materials and foodstuffs and an importer of finished goods. As can

**Table 3.1: Composition of U.S. Exports, 1866–1914 (Per Cent)**

| *Years* | *Primary Goods* | *Finished Goods*[a] |
|---|---|---|
| 1866–70 | 67 | 33 |
| 1871–75 | 60 | 40 |
| 1876–80 | 56 | 44 |
| 1881–85 | 55 | 45 |
| 1886–90 | 53 | 47 |
| 1891–95 | 51 | 49 |
| 1896–1900 | 45 | 55 |
| 1901–05 | 42 | 58 |
| 1906–10 | 41 | 59 |
| 1911–14 | 39 | 61 |

a. Includes semi-manufactures.

Calculated from: *Historical Statistics of the United States, Colonial Times to 1957: A Statistical Abstract Supplement* (U.S. Government Printing Office, Washington, 1960), pp. 544–45.

be seen in Table 3.1, U.S. exports of finished goods did not exceed those of raw materials until the 1890s. 1893, in fact, was the first year in which the value of finished goods exports was greater than that of unfinished goods.[45]

Refined oil, on the other hand, became a leading export item long before 1893. For many years it was the second most important manufactured export, exceeded only by meat products.[46] As Table 3.2 shows, petroleum products accounted for a significant share of finished goods exports in the pre–World War I years. This share ranged from a high of 15 per cent in the late 1880s to a low of 10 per cent in the early 1900s.

This trade in oil naturally aroused the interest of U.S. government leaders, who were intent on defending and expanding export markets. U.S. diplomatic representatives had played an important role in promoting sales of U.S. oil to foreign countries since the early 1860s.[47] One of the most overt moves backing oil exports was the declaration in the late 1890s of the "open door" policy towards China. While the U.S. secretary of state, John Hay, enunciated this policy, designed to open up China's markets to U.S. interests, mainly in response to pressure from exporters of cotton and from financial groups wanting to build railroads in China,[48] there is little doubt that he was also interested in expanding U.S. oil markets there. As can be seen in Table 3.3, the Asian market as a whole, and the Chinese market in particular, were becoming increasingly important to the U.S. oil industry. These markets were replacing the losses the industry was suffering in Europe following the rise of Russian oil. By the turn of the century Asia and China were the fastest-growing markets for American oil. These growing markets were,

**Table 3.2: Share of Petroleum Products in U.S. Finished Goods Exports, 1886–1914 (Per Cent)**

| *Years* | *Petroleum Products* |
|---|---|
| 1911–14 | 10 |
| 1906–10 | 10 |
| 1901–05 | 10 |
| 1896–1900 | 11 |
| 1891–95 | 11 |
| 1886–90 | 15 |

Calculated from: *Historical Statistics of the United States, Colonial Times to 1957: A Statistical Abstract Supplement* (U.S. Government Printing Office, Washington, 1960), pp. 544, 546.

**Table 3.3: Geographic Distribution of U.S. Exports of Illuminating Oil, 1874 and 1899 (Per Cent)**

| *Geographic Area* | *1874* | *1899* | *Per Cent Change, 1874–99* |
|---|---|---|---|
| Europe | 87 | 63 | -20 |
| Asia | 4 | 20 | 473 |
| (China) | (1) | (4) | (814) |
| South America | 3 | 7 | 187 |
| North America | 3 | 3 | 25 |
| Other | 3 | 7 | 70 |
| Total | 100 | 100 | 9 |

Calculated from: H. Williamson & A. Daum, *The American Petroleum Industry, The Age of Illumination 1859–1899* (Northwestern University Press, Evanston, Ill., 1959), pp. 742–43, 752.

however, being threatened by the French Rothschilds, who were marketing Sumatran oil there.[49] The "open door" policy unquestionably served to buttress the U.S. oil position in China *vis-à-vis* its competition.

The prime beneficiary of U.S. oil policy was, of course, Rockefeller, who had a virtual monopoly in the export trade. His only significant U.S. competitor was Pure Oil Company, which had only 5 per cent of the market.[50] The benefits of the "open door" policy to Rockefeller were quite substantial. Standard Oil's share of the Asian market in illuminating oil rose from 34 per cent in 1899 to 46 per cent in 1908.[51] Rockefeller, in his *Random Reminiscences*, openly acknowledged the aid rendered him by the U.S. government. "One of our greatest helpers," he wrote, "has been the State Department in Washington. Our ambassadors and ministers and consuls have aided to push our way into new markets to the utmost corners of the world."[52] This close link between the U.S. State Department and the Rockefeller-controlled oil companies has, as we shall see, continued to the present day. What is especially interesting is that this link remained firm even in those periods when the U.S. government was prosecuting these companies for their domestic activities.

## Foreign Investment

This aid of the U.S. government was important to Rockefeller in meeting his competition abroad where, unlike in the U.S., he encountered major competitors with enormous reserves of crude. Rockefeller tried to counter this competition through the acquisition of refining and distribution facilities abroad. Thus, at a relatively early stage in its history, Standard Oil began to make direct investments overseas.

It is interesting to note that Standard initially made market-oriented, rather than supply-oriented, investments. This followed the pattern in the United States, a pattern that enabled him to achieve a dominant position in the domestic market. Furthermore, believing he could triumph over his foreign competitors, Rockefeller originally eschewed making any market-sharing arrangements with them. His subsequent failure to destroy his competitors forced him both to acquire producing properties and to enter into cartel agreements.

The first major foreign investment by Standard Oil was the establishment in 1888 of the Anglo-American Oil Co. to market U.S. refined products in England.[53] Several years later he formed the Deutsch-Amerikanische Petroleum Gessellschaft (DAPG) to perform the same function in Germany.[54] Rockefeller used DAPG not only to pre-empt the German market from Russian oil but also to pre-empt it from his U.S. competitors. In 1896, six years after its incorporation, DAPG bought out Philip Poth, its main competitor in Germany.[55] Poth had been distributing oil supplied by the Pure Oil Company.[56]

Around the same time Standard set up a French affiliate, Bedford et Compagnie, which purchased an existing refinery and began construction of a new one. This move aroused strong opposition within the French refining industry. The dispute was resolved through a cartel agreement between Rockefeller and the French refiners. The accord provided that all French companies would buy the bulk of their crude from Standard. In return Standard agreed not to sell most of its refined products to anyone outside the French group.[57] Thus Rockefeller began using in Europe the same exclusionary tactics that had made him so powerful in the United States.

Standard at various times either purchased or built refineries in Cuba,[58] Puerto Rico (when it was still under Spanish rule),[59] Germany,[60] Austro-Hungary, Japan, and Canada.[61] The acquisition of refineries in Canada came about through the purchase of controlling interest in Imperial Oil, Canada's largest oil company, in 1898.

Imperial had been organized in 1880 by sixteen independent refiners in London, Ontario, who formed the company to protect themselves against Standard Oil.[62] Led by a transplanted Clevelander, Jacob Engelhart, Imperial rationalized its operations by closing down inefficient refineries and concentrating production in the more efficient ones.

Engelhart also built new refineries, and improved the quality of the product by hiring Herman Frasch (who later developed the Frasch process for producing sulphur from deep mines) to find a way of reducing the sulphur content of the kerosine. By 1883 Imperial had developed twenty-three branch offices from coast-to-coast and exported kerosine to China, India, Japan, Australia, and South America.[63]

Despite Imperial's successes, it was pressed hard by Standard Oil, which, by the late 1890s, had acquired one-third of the Canadian market. To stay ahead of Rockefeller, Imperial required large amounts of capital. Engelhart searched for capital in Canada and in England but was unable to raise sufficient funds. Finally, he went to the only possible source of capital, Standard Oil, which gave him the capital in return for a controlling interest in Imperial.[64] Through this transaction, Standard not only eliminated Imperial as a competitor, but also obtained a fully integrated company which had producing properties as well as refining, transportation, and marketing facilities.[65]

Through control of Imperial, Standard achieved a dominant position in the Canadian market. It also used Imperial as a vehicle for investing in those countries where Americans were not welcome. When, on the eve of World War I, Standard decided to acquire producing properties abroad, it used Imperial to obtain the controversial La Brea y Parinas fields in Peru. Through Imperial, Standard organized the International Petroleum Company (IPC), ostensibly a Canadian company, to run its fields and refineries in Peru, where IPC subsequently became the leading oil company.[66]

Standard used another affiliate, Waters-Pierce Oil Company, to penetrate the Mexican market. Waters-Pierce, Standard's exclusive marketing firm in the U.S. South-West,[67] built three Mexican refineries, which purchased all their crude from Standard,[68] and acquired producing properties in Mexico.[69] Standard used its Dutch marketing affiliate, the American Petroleum Company, to acquire concessions in Indonesia in 1912.[70] It used the affiliate route of acquisition because in 1899 the Dutch government had blocked its direct attempts to obtain concessions there.[71]

The British government, too, stopped it from investing in Burma's oil fields in 1902.[72] Standard did not attempt to circumvent this prohibition. Rumania, on the other hand, had no objection to Standard acquiring producing properties within its boundaries. Standard acquired 100 per cent ownership of Romano-American Company, a fully integrated company with both refineries and oil fields.[73]

All told, before its dissolution in 1911, Standard had sixty-seven

**Table 3.4: U.S. Share of World Petroleum Market Outside the United States, 1899–1914 (Per Cent)**

| *Year* | *Share* | *Year* | *Share* |
|---|---|---|---|
| 1899 | 23 | 1907 | 25 |
| 1900 | 21 | 1908 | 27 |
| 1901 | 21 | 1909 | 25 |
| 1902 | 21 | 1910 | 24 |
| 1903 | 19 | 1911 | 26 |
| 1904 | 19 | 1912 | 27 |
| 1905 | 27 | 1913 | 30 |
| 1906 | 27 | 1914 | 31 |

Calculated from: United States Senate, *Petroleum Interests in Foreign Countries Hearings before a Special Committee Investigating Petroleum Resources* (U.S. Government Printing Office, Washington, 1946), p. 214.

affiliates engaged in foreign trade.[74] By the end of 1907, it controlled fifty-five foreign companies with a capitalization of approximately $37 million.[75] Nevertheless Rockefeller was never able to achieve the dominance in foreign markets that he attained in the domestic one. In 1910, one year before dissolution, Standard Oil's share of overseas crude production was less than 1 per cent.[76] As far as products were concerned, Standard's share of the overseas petroleum market never reached one-third in the years before World War I. As indicated in Table 3.4, the share of all U.S. companies in the foreign market fell from approximately one-quarter to one-fifth in the early years of the century, and then rose to approximately 30 per cent by 1914.

In the years before the war several other U.S. companies entered the foreign field. They included Pure Oil, which established its own marketing facilities in Europe after DAPG acquired Philip Poth.[77] These, however, were subsequently sold to Shell and Standard.[78] The Texas Company began investing in sales outlets throughout the world. It also invested in producing properties in Mexico, as did Gulf, Sinclair, and a host of other American companies.[79]

This overall growth of direct foreign investments by the petroleum industry was, as was indicated in the preceding chapter, part and parcel of the general overseas expansion of American business in the pre–World War I era. There was nothing particularly spectacular about the growth of the oil industry beyond the U.S. borders. Its growth was, in fact, slower than the growth of total direct foreign investment, as can be seen in Table 3.5.

**Table 3.5: Per Cent Changes in U.S. Direct Foreign Investment by Industry, 1897–1914**

| *Industry* | *Change* |
|---|---|
| Other[a] | 796 |
| Utilities | 504 |
| Mining and Smelting | 437 |
| Manufacturing | 408 |
| Agriculture and Timber | 362 |
| Petroleum[b] | 304 |
| Sales Organizations[c] | 198 |
| Railroads[d] | 77 |
| Total | 318 |

a. Mainly banks and insurance companies.
b. Includes exploration, production, refining, and distribution. Excludes investments in Indonesia.
c. Includes trading companies and sales branches and subsidiaries of large corporations. Excludes petroleum distribution.
d. Includes U.S. government purchase of Panama Canal Railroad.

Calculated from: C. Lewis, *America's Stake in International Investment* (The Brookings Institution, Washington, 1938), pp. 579, 588, and M. Wilkins, *The Emergence of Multinational Enterprise, American Business Abroad from the Colonial Era to 1914* (Harvard University Press, Cambridge, 1970), p. 110.

There were several reasons for this relatively slow growth. Firstly, the growing domestic market for oil, especially for use in automobiles, tended to keep investment at home. Secondly, many foreign governments discouraged the entry of Rockefeller-owned companies into their countries. Thirdly, Rockefeller's policy of investing in refining and marketing, rather than in extraction, restricted investments to those countries where living standards were relatively high. Finally, and probably most important, the U.S. government made no special effort to promote oil investment abroad. Though the government did encourage and help oil companies invest overseas, it did so as part of a general policy to spread American business abroad. It saw no reason why it should give preferential treatment to the oil companies.

The events of World War I dramatically changed the U.S. government's attitude. The Great War demonstrated the importance of oil in modern warfare. From that time on, obtaining control of oil became one of the cornerstones of U.S. foreign economic policy. It became the essential ingredient in the expansion and consolidation of the American Empire.

## Notes

1. "Colonel" Drake was in reality a railroad conductor. The New England promoters backing his project addressed letters to him with the title of Colonel in order to impress the residents of Titusville. This type of promotional activity was not untypical of future oil industry behaviour. See H. F. Williamson & A. R. Daum, *The American Petroleum Industry, The Age of Illumination* (Northwestern University Press, Evanston, Ill., 1959), pp. 75–76.
2. H. O'Connor, *World Crisis in Oil* (Monthly Review Press, New York, 1962), p. 27.
3. In 1857, two years before the successful completion of Drake's well, James Williams, a Canadian businessman, used a combination of digging and drilling to obtain oil in Ontario. Drake, however, concentrated entirely on drilling. See E. Gray, *Impact of Oil, The Development of Canada's Oil Resources* (The Ryerson Press/Maclean-Hunter, Toronto, 1969), pp. 5–6.
4. Williamson & Daum, *The American Petroleum Industry*, p. 40.
5. One of the leaders in developing the refining process was Dr. Abraham Gesner, a Canadian physician. See ibid., pp. 44–48.
6. Gray, *The Impact of Oil*, p. 7.
7. Ibid., p. 8.
8. J. Schumpeter, *Capitalism, Socialism, and Democracy*, 2nd ed. (Harper and Brothers, New York, 1947), chap. 7.
9. Williamson & Daum, *The American Petroleum Industry*, pp. 483–84.
10. Ibid., p. 483.
11. Ibid., p. 471.
12. Ibid., pp. 320, 325.
13. Ibid., p. 473.
14. Calculated from ibid., pp. 326, 349, 375.
15. Ibid., p. 567.
16. Schumpeter, *Capitalism, Socialism and Democracy*, p. 83.
17. A rebate is a refund to the shipper on his own shipments. A drawback is a payment to one shipper on the shipments of others. Thus the payments Rockefeller's competitors made to the railroads included not only the regular shipping charges but payments to Rockefeller as well.
18. I. Tarbell, *The History of the Standard Oil Company* (2 vols., The Macmillan Company, New York, 1933), vol. 1, p. 63.
19. Ibid., pp. 67–68.
20. Williamson & Daum, *The American Petroleum Industry*, p. 474.
21. H. Williamson, R. Andreano, A. Daum & G. Klose, *The American Petroleum Industry, The Age of Energy, 1899-1959* (Northwestern University Press, Evanston, Ill., 1963), p. 7.
22. G. Montague, *The Rise and Progress of the Standard Oil Company* (Harper and Brothers, New York, 1903), p. 131.
23. Williamson & Daum, *The American Petroleum Industry*, pp. 564–68.
24. Williamson, Andreano, Daum & Klose, *The American Petroleum Industry*, p. 7.
25. I. Tarbell, *The History of The Standard Oil Company*, vol. 2, pp. 40–41.

26. Ibid., p. 43.
27. P. Collier & D. Horowitz, *The Rockefellers: An American Dynasty* (New American Library, New York, 1976), p. 54.
28. R. O'Connor, *The Oil Barons: Men of Greed and Grandeur* (Little, Brown and Company, Boston, 1971), p. 59.
29. Collier & Horowitz, *The Rockefellers*, p. 4.
30. *Standard Oil Co. of New Jersey* v. *United States*, 221 U.S. 1 (1911).
31. *United States* v. *United States Steel Corporation et al.*, 251 U.S. 417 (1920).
32. Ibid.
33. *Standard Oil Co. of New Jersey* v. *United States.*
34. Collier & Horowitz, *The Rockefellers*, p. 58.
35. Williamson & Daum, *The American Petroleum Industry*, p. 605.
36. Ibid., p. 607
37. *Standard Oil Co. of New Jersey* v. *United States.*
38. Lucas was the Anglicized version of Luchich, his family name in the place of his birth, then the province of Dalmatia in the Austro-Hungarian Empire and now Yugoslavia. He preferred the Anglicized version. See H. O'Connor, *The Empire of Oil* (Monthly Review Press, New York, 1962), p. 8.
39. R. O'Connor, *The Oil Barons*, p. 71.
40. C. Solberg, *Oil Power* (Mason/Charter, New York, 1976), pp. 53–54.
41. Ibid., p. 59.
42. Williamson, Andreano, Daum & Klose, *The American Petroleum Industry*, p. 7.
43. Williamson & Daum, *The American Petroleum Industry*, p. 302.
44. Calculated from ibid., pp. 737–38.
45. *Historical Statistics of the United States, Colonial Times to 1957: A Statistical Abstract Supplement* (U.S. Government Printing Office, Washington, 1960), p. 544.
46. Ibid., p. 546.
47. Williamson & Daum, *The American Petroleum Industry*, pp. 324, 493.
48. C. S. Campbell, Jr., "American Business Interests and the Open Door in China," *The Far Eastern Quarterly*, vol. 1, no. 1 (1941), pp. 43–44.
49. Williamson, Andreano, Daum & Klose, *The American Petroleum Industry*, p. 258.
50. Ibid., p. 251.
51. Ibid., p. 257.
52. Collier & Horowitz, *The Rockefellers*, p. 40.
53. M. Wilkins, *The Emergence of Multinational Enterprise: American Business Abroad from the Colonial Era to 1914* (Harvard University Press, Cambridge, 1970), p. 64.
54. Williamson & Daum, *The American Petroleum Industry*, p. 648.
55. Ibid., p. 653.
56. Ibid., p. 569.
57. Ibid., p. 652.
58. Ibid., p. 661.
59. Ibid.
60. C. Lewis, *America's Stake in International Investment* (The Brookings Institution, Washington, 1938), p. 581.

61. Ibid., p. 582.
62. E. Gray, *The Great Canadian Oil Patch* (Maclean-Hunter Limited, Toronto, 1970), pp. 258, 261.
63. Ibid., pp. 261–64.
64. Ibid., p. 264.
65. Wilkins, *The Emergence of Multinational Enterprise*, p. 140.
66. Ibid., p. 186.
67. Williamson & Daum, *The American Petroleum Industry*, p. 543.
68. Lewis, *America's Stake in International Investments*, p. 589.
69. Wilkins, *The Emergence of Multinational Enterprise*, p. 85.
70. Ibid., pp. 85–86.
71. Ibid., p. 84.
72. Ibid.
73. Ibid.
74. Ibid.
75. Ibid., p. 83.
76. Williamson, Andreano, Daum & Klose, *The American Petroleum Industry*, p. 258.
77. Williamson & Daum, *The American Petroleum Industry*, p. 577.
78. Wilkins, *The Emergence of Multinational Enterprise*, p. 86.
79. Ibid. and Lewis, *America's Stake in International Investments*, p. 589.

CHAPTER FOUR

# The Relationship between the Oil Industry and the American State—The Interwar Period

In actuality the British, rather than the Americans, were the first to recognize the military potential of oil. As early as 1899, Marcus Samuel, the head of Shell, began campaigning to convert the British navy from coal to oil.[1] Initially the admirals resisted Samuel's suggestion, suspecting that it was motivated by self-interest. They also had doubts about the technical feasibility of oil.

## Churchill's Conversion to Oil

Nevertheless, Samuel managed to convince the first sea lord, Admiral Fisher, of the soundness of his ideas. Whether Samuel's offer of a Shell directorship to Fisher played a role in his conversion is still a matter of conjecture.[2] In any event Fisher waged a vigorous campaign to transform the British navy into an oil-burning fleet. He found an ally in Winston Churchill, who became first lord of the admiralty in 1911.

Fearing the growing strength of the German navy, Churchill wanted to assure the supremacy of the British fleet in the war that he knew was sure to come. In 1912 he established The Royal Commission on Oil Supplies, headed by Lord Fisher, to examine the advantages of oil.[3] Not surprisingly, the commission came out in favour of oil. This decision turned out to be a sound one as it did give the British navy a distinct advantage over the German navy in World War I. In later years Churchill outlined the superiority of oil.

> In equal ships, oil gave a large excess of speed over coal. It enabled their speed to be attained with far greater rapidity. It gave 40 per cent greater radius of action for the same rate of coal. It enabled the fleet to be refueled at sea with facility. . . . The use of oil made it possible in every type of vessel to have more gun-power and more speed for less size and less cost.[4]

Though the Royal Commission showed that the technical objections to oil were groundless, it still had to deal with the problem of security of supply. Britain had ample coal but practically no oil. Unless secure oil supplies were obtained, the British fleet could be rendered inoperative in a conflict.

The most obvious supplier of oil was Shell. Marcus Samuel, a solid member of the British establishment, had, after all, initiated the idea. But there was strong opposition to Shell among members of the commission and in Parliament. Shell had increased the price of oil, arousing a great deal of popular resentment. Sir Marcus (he had been knighted in the early 1900s)[5] defended the increase, stating that "the price of an article is exactly what it will fetch."[6] This attitude caused Churchill to remark:

> It is their policy. . . to acquire control of the sources and means of supply, and then to regulate the production and the market price. . . . We have no quarrel with Shell. We have always found them courteous, considerate, ready to oblige, anxious to serve the Admiralty and to promote the British Navy and the British Empire—at a price.[7]

But perhaps more fundamental was the fear that Shell was not under British control. Several years earlier Rockefeller had attempted to buy out Shell. Though Samuel rejected this offer,[8] he merged his company with Henry Deterding's Royal Dutch Petroleum to form Royal Dutch Shell.[9] Under the terms of this 1906 agreement, the Dutch interests obtained 60 per cent of the shares with British interests holding the remainder.

Churchill actually regarded Shell as a Dutch company, and suspected that Deterding was under German influence.[10] He did not want such a company to be the main supplier of the British navy. As it turned out, Churchill's fears were ill-founded. During the war the company, despite the proclaimed neutrality of Holland, sided with Britain. It did not overcharge the British navy, even though it was in an excellent position to do so.[11]

Another generally held objection was that none of Shell's producing properties was in the British Empire. Shell's properties lay in Borneo and in the Dutch East Indies. The British could not be sure that these supplies would be available to them in a war. They wanted to be supplied by a company with reserves within the Empire. The only such company in existence was Burmah Oil, which had been producing oil in Burma since 1886[12] and which was, furthermore, entirely controlled by British interests.

By 1912, however, Burmah's real wealth lay not in Burma but in Persia. In 1901 William D'Arcy, a British engineer, had obtained a concession covering 500,000 square miles from the Persian government. Unsuccessful in his early attempts to find oil, D'Arcy was on the verge of bankruptcy when the British government asked Burmah Oil to bail him out. In 1905 Burmah founded the Concessions Syndicate to handle its Persian venture. When oil was finally discovered in 1908, Concessions Syndicate formed the Anglo-Persian Oil Company,[13] which changed its name in 1935 to Anglo-Iranian, then in 1954 to British Petroleum, the name by which it is presently known.[14]

Though technically this Persian oil was outside the British Empire, it was well within the British sphere of influence and could be defended from bases in India. In fact, in 1905 the British had sent a gunboat and a detachment of Bengal Lancers to Persia to protect the company's properties against attacks by marauding tribesmen.[15] Thus, the Anglo-Persian Oil Company, controlled by British interests and with reserves within the British sphere of influence, was looked upon as the ideal company to supply the Royal Navy.

By 1913, the time that the Royal Commission concluded its hearings, this company was having financial difficulties. Though it had invested large sums in Persia, it still had not sold a single barrel of oil.[16] There was a danger that this troubled company might fall into the hands of foreign interests, especially those of Rockefeller's Standard Oil. Churchill was determined that this should not happen. He therefore proposed to a somewhat shocked Parliament that the British government buy a controlling interest in Anglo-Persian to ensure that it would always remain in British hands.

> We must become the owners, or at any rate the controllers at the source, of at least a proportion of the supply of natural oil which we require . . . and obtain our oil supply, so far as possible, from sources under British control, or British influence.[17]

After considerable debate, intermingled with accusations of "socialism," Parliament in 1914 approved the purchase by an overwhelming vote of 254 to 18.[18] The government paid £2 million for a 51 per cent share in the company. In return it imposed a stipulation that Anglo-Persian must always remain a British concern and that every director must be a British subject. Though the government acquired the majority of shares, it obtained the right to appoint only two of the company's seven directors. Furthermore, the government assured the companies that these directors, who had the right of veto, would exercise that right only in matters involving foreign or military policy or admiralty contracts.[19] This is one promise that the British kept. Neither that government nor any subsequent one has ever exercised the right of veto.[20] In other words the British permitted the private interests to have free rein with public funds.

This marriage of private and public interests marked a new era in the development of British capitalism. As Hartshorn has remarked:

> What is germane in this context is that the decision to buy control of this company represented the first open manifestation of the vital interest to the government of a country lacking indigenous oil supplies of securing a source of supply that it recognized as strategically vital. But it was the prototype. Britain sought and gained command of secured "tied" supplies of oil from a promising source that later turned out to be part of the richest oil-bearing region yet discovered. Over the years it extended that command in the region . . .[21]

## Oil Imperialism

What Hartshorn described is the policy of oil imperialism adopted by Churchill, *i.e.*, a deliberate policy of acquiring oil reserves in foreign lands. No longer was Britain, the traditional bastion of free trade, content to rely on the market place to provide it with this vital commodity. In its quest for oil it abandoned *laisser-faire* capitalism in favour of a state-owned company buttressed by the full might of the empire.

It was inevitable that other nations would follow Britain's example. World War I underscored the strategic importance of oil. Oil was used to power not only the ships but also the new instruments of warfare: the tank, the airplane, and the truck. President Clemenceau of France recognized this new reality when he wrote to President Wilson of the U.S. that oil was "as necessary as blood."[22] He was echoed by his oil commissioner, Henri Berenguer, who proclaimed that he "who has oil has Empire" and that oil "is more precious, more penetrating, more influential in the world than gold itself."[23] Not surprisingly, the French government followed a policy similar to Britain's after the war had ended. It bought shares in Compagnie Francaise des Petroles (CFP) in order to guarantee for itself a portion of that company's reserves in Iraq.[24]

The war also brought home to the United States the significance of oil. That country furnished 80 per cent of the Allies' petroleum needs.[25] In order to do so and to supply its armed forces as well, it imported oil from Mexico[26] and instituted a program of "gasless Sundays" in which people voluntarily refrained from using their automobiles except for "errands of necessity."[27]

These war-time shortages gave rise to fears that the United States might not have sufficient oil reserves to see it through another war. Oil thus began to be a matter of special concern to U.S. policy-makers. Mark Requa, director of the oil division of the U.S. Fuel Administration, advocated a policy of encouraging "in every way possible. . .the acquisitions by our nationals of reserves in foreign lands."[28] These foreign reserves were to be considered part of the U.S. oil supply.

It is interesting to note that U.S. oil policy differed somewhat from the policies of England and France. The latter countries used companies in which the governments had invested, while the United States relied wholly on privately-owned companies to achieve the same end—the attainment of a secure source of oil.[29] This difference stemmed from the changes in international investment wrought by the war. In 1914, at the beginning of the war, the U.S. was a net debtor, owing the rest of the world $3.7 billion. By 1919 it had shifted to a net creditor position with claims on the rest of the world of $3.7 billion.[30] The U.S. had become the main supplier of capital to the world. There was little danger that any large American firms would be bought out by foreigners, and thus there was no need for the government to assume an equity position in any company to keep it under domestic control. Besides, by the end of the war there were a large number of U.S. oil firms in existence. If the government wanted to assure domestic control through assuming equity positions, it would either have to (1) buy control of all of them or (2) select only a limited number of them for this purpose.

The first alternative would have been quite costly. In addition, it would have been viewed as a step towards the nationalization of the oil industry, a step which would have been politically unacceptable. The second alternative would have involved the government in the politically embarrassing choice of selecting which companies to purchase. Furthermore, it would have placed these government-controlled companies in competition with the privately-controlled ones, again leading to political difficulties. Therefore, the government, instead of selecting a single or small group of companies to use as its agent, assigned that role to all private oil companies operating abroad. By the end of the war the U.S. policy had shifted from one of aiding the overseas expansion of all U.S. firms, the oil companies among them, to one of giving top priority to oil. Oil thenceforth was to receive special treatment in U.S. foreign economic policy. The U.S. had formally entered the era of oil imperialism.

In pursuing its objectives, the U.S. adopted two distinct and contradictory doctrines: "minimum duty" and the "open door."

*Minimum Duty*

The doctrine of "minimum duty" emerged in the argument with Mexico over that country's attempt in 1918 to nationalize sub-surface mineral rights. During the latter part of the nineteenth century Porfirio Diaz, the president of Mexico, granted large concessions to railroads, mines, and agricultural enterprises. These concessions included ownership of sub-surface resources. Since in most parts of the world outside the United States these resources were considered the property of the state, many Mexicans questioned the legality of these grants.

Among those receiving such a concession was Edward L. Doheny, an American. In 1901 Doheny discovered oil on his land and Mexico entered the oil era. Oil companies from the United States and England swarmed into Mexico seeking concessions from the government. Behind the scenes a bitter struggle ensued between the U.S. and British governments over control of Mexico. Diaz, who did not want to be dominated by any single power, welcomed the British as a counterweight to the Americans. Diaz's pro-British policies caused concern among American investors in Mexico. In 1910, shortly after he was elected president for the eighth time, he was ousted in a revolution led by Francisco Madero. Subsequent testimony before the U.S. Senate Foreign Relations Committee indicated that Madero received funds from "several leading bankers in El Paso" and from Rockefeller's Standard Oil.[31] There was no evidence of direct U.S. government involvement in the ouster. Nevertheless, Nearing and Freeman noted that "the sentiment in the United States was strongly in favour of Madero and the United States Government took the earliest possible opportunity of recognizing him."[32]

The United States, on the other hand, did not extend recognition to Victoriano Huerta, who overthrew Madero in 1913. Huerta was financed by English interests.[33] Britain's support of Huerta infuriated the U.S. government. Lawrence, in his biography of President Wilson, stated:

> Mr. Wilson had many tempestuous conflicts with the British Foreign Office over the apparent support given to the Huerta regime by Sir Lionel Carden, the British Minister to Mexico, a support intensified to no small extent by the large British oil companies in Mexico whose influence in London circles was appreciable.[34]

Wilson regarded this British support as an obstacle to his attempts to overthrow Huerta. After much wrangling Wilson induced the British to withdraw their support, in return for which the U.S. agreed to abolish tolls on shipments through the Panama Canal.[35] After this agreement the U.S. government moved against the Huerta regime. It openly sold arms to Venustiano Carranza and Francisco Villa, who were leading a rebellion against the government. The U.S. also sent troops into the port city of Vera Cruz to prevent the Huerta forces from receiving arms and to deprive Huerta of much-needed customs revenues.[36] Through these tactics the U.S. forced Huerta from office and installed Carranza as president on the eve of World War I.[37]

Carranza's forces also received financial support from private U.S. oil interests, like Doheny[38] and Pierce,[39] of the Rockefeller-controlled Waters-Pierce Oil Company. They had good reason to back a political leader who presumably favoured their interests. These and other American interests had a huge stake in Mexican oil. As can be seen in Table 4.1, U.S. companies had investments of $85 million in production facilities in 1914. Almost three-fifths of oil production investments were

**Table 4.1: Geographic Distribution of U.S. Direct Foreign Investment in Oil Production,[a] 1914 and 1919**

| *Country* | *1914* Investment ($ Millions) | *1914* (%) | *1919* Investment ($ Millions) | *1919* (%) |
|---|---|---|---|---|
| Western Hemisphere | 135 | 94[b] | 321 | 98 |
| Mexico | (85) | (59) | (200) | (61) |
| Canada | (25) | (17) | (30) | ( 9) |
| Peru | (15) | (10) | (45) | (14) |
| Venezuela | ( 5) | ( 3) | (18) | ( 5) |
| Trinidad | ( 3) | ( 2) | ( 5) | ( 2) |
| Colombia | ( 2) | ( 1) | (20) | ( 6) |
| Central America | ( 0) | ( 0) | ( 3) | ( 1) |
| Eastern Hemisphere | 8 | 6 | 7.5 | 2 |
| Rumania | ( 5) | ( 3) | ( 7.5) | ( 2) |
| Russia | ( 3) | ( 2) | ( 0) | ( 0) |
| Totals | 143 | 100 | 328.5 | 100 |

a. Includes refining.

b. Because of rounding the total may not equal the sum of the column.

Source: C. Lewis, *America's Stake in International Investment* (Brookings Institution, Washington, 1938), p. 588.

in Mexico. By 1914 Mexico had become a major producer. Its oil output of 26 million barrels was exceeded only by that of the United States and Russia.[40] Approximately two-thirds of this quantity was exported to the United States.[41]

In the years following Carranza's accession to power U.S. oil investments in Mexico, Mexican production, and Mexican exports to the U.S. continued to grow. By 1919, as Table 4.1 shows, investments had risen to $200 million. The U.S. firms in Mexico, in addition to those controlled by Doheny and Pierce, were Gulf, Sinclair, Atlantic, and Standard Oil (N.J.).[42] As Table 4.2 indicates, Mexican production skyrocketed during those years, rising from 26 million barrels in 1914 to 87 million in 1919. By the latter year Mexico replaced Russia as the world's second largest producer.[43] In that year Mexico produced about one-sixth of the earth's oil. Its rise in production was greater than any other country with the exception of Persia, where production expanded sharply to meet the war-time needs of the British navy, nevertheless remaining relatively small, only 10 million barrels.

During this period Mexico's exports to the U.S. rose from 18 million to 53 million barrels or by 200 per cent. By 1919 these exports equalled 14 per cent of U.S. production.[44] By any measure Mexican oil was very important both to the U.S. oil industry and to the U.S. economy.

**Table 4.2: World Production of Crude Oil, 1914 and 1919**

| *Country* | *1914* Production (Millions of Barrels) | *(%)* | *1919* Production (Millions of Barrels) | *(%)* | *Per Cent Change 1914–19* |
|---|---|---|---|---|---|
| World | 407.5 | 100 | 555.9 | 100 | 36 |
| Major Producing Areas | 392.5 | 96 | 535.4 | 96[a] | 36 |
| United States | (265.8) | (65) | (378.4) | (68) | 42 |
| Russia | ( 67.0) | (16) | ( 31.7) | ( 6) | –53 |
| Mexico | ( 26.2) | ( 6) | ( 87.1) | (16) | 232 |
| Rumania | ( 12.8) | ( 3) | ( 6.6) | ( 1) | –48 |
| Dutch East Indies | ( 11.4) | ( 3) | ( 15.5) | ( 3) | 36 |
| Poland | ( 6.4) | ( 2) | ( 6.0) | ( 1) | –6 |
| Persia | ( 2.9) | ( 1) | ( 10.1) | ( 2) | 249 |

a. Because of rounding the total may not equal the sum of the column.

Source: H. Williamson, R. Andreano, A. Daum & G. Klose, *The American Petroleum Industry, The Age of Energy, 1899–1959* (Northwestern University Press, Evanston, Ill., 1963), p. 262.

It is therefore not surprising that the U.S. government reacted with extreme hostility towards Mexico when Carranza, the man whom it put in power, moved against the oil industry. Carranza proposed a new constitution, which would have nationalized sub-surface mineral resources. This was a direct blow to those oil companies that had received concessions but had not commenced drilling over the entire area of their concessions. The great majority of companies were in this category. The Mexican government generally refused to issue drilling permits to these companies unless they acknowledged the nation's ownership of the subsurface resources.

The U.S. government protested these actions, claiming that they were retroactive. Mexico, for its part, denied the retroactivity charge, contending that the nation was merely reclaiming the resources that had been illegally ceded to the companies by Diaz. The Mexican government also pointed out that, as a sovereign nation, it had the right to dispose of its resources in any way it saw fit.

In reply the United States set forth its doctrine of "minimum duty" in a note sent to the Mexican government in 1918. Henry Fraser, chief counsel to the U.S. Senate Special Committee Investigating Petroleum Resources, later summarized the U.S. position as follows:

> The principal point made by the United States. . . was that every nation has certain minimum duties to perform with regard to the treatment of foreigners, irrespective of its duties to its own citizens, and that in default of such performance it is the right of the foreign government to enter protest. Since one of the duties imposed upon a government is to refrain from measures resulting in confiscation of the vested property rights of foreigners, the United States is not estopped, the note said, from protesting on behalf of its citizens merely because the Mexican Government may choose to confiscate vested property rights of its own citizens. The note concluded with the hope that the Mexican courts would protect the legitimately acquired rights of American citizens and thus happily end the controversy; but if this hope should be disappointed, "the Government of the United States must reserve to itself the consideration of the questions of interesting itself on behalf of American citizens concerned in this important and serious matter."[45]

There is no question that this note threatened armed intervention unless the Mexican government adopted the doctrine of "minimum duty." The Mexican government under President Calles finally acceded to the U.S. demands in 1928.[46] This U.S. victory was relatively short-lived, as Mexico finally nationalized the oil properties in 1938.[47] The implications of this nationalization will be discussed later. What is important at this point are the implications of "minimum duty." This doctrine in effect said that foreigners—or, more precisely, foreign investors—are a special, privileged class, not subject to the laws of the nation in which they do business. Any nation could enact any legislation it wished regulating the business activities of its citizens as long as these laws did not apply to foreigners.

This doctrine served the convenience of the United States in its dealings with Mexico. It did not, however, serve the interests of the United States in its dealings with Britain and Holland in the post–World War I period. Against them, the U.S. used the doctrine of the "open door." That these two doctrines conflicted with each other never seemed to bother U.S. policy-makers. They used whatever doctrine served their purpose at the moment. The doctrine of "minimum duty," with its provision of special privilege, was suitable for a country in which the U.S. had already acquired a large vested interest. The doctrine of "open door," with its implications of non-discrimination, was, on the other hand, suitable for an area from which U.S. interests were being excluded. In any event both doctrines served the goal of protecting and expanding U.S. oil interests abroad.

### *Open Door*

#### IRAQ

The U.S. first used the "open door" doctrine as a means of forcing the British to give U.S. oil companies the right to participate in the exploitation of Iraq's oil resources. The British, after long and often acrimonious discussions, finally agreed to let the American companies in.

The dispute centred on the postwar reorganization of the Turkish Petroleum Company (TPC). This company, formed in 1912 when Turkey held suzerainty over Iraq, represented a combination of British and German interests put together by C. S. Gulbenkian, a Turkish Armenian, to exploit the petroleum resources of the provinces of Mosul and Bagdad. The original shareholders of the company were the National Bank of Turkey, which was owned entirely by British interests; Anglo-Saxon Petroleum company, a subsidiary of Royal Dutch Shell; and the Deutsche Bank. The shares were divided as follows: 25 per cent each for Anglo-Saxon and the Deutsche Bank and 50 per cent for the National Bank of Turkey, of which 15 per cent was held for Gulbenkian.[48] Early in 1914 the National Bank's holdings were transferred to Anglo-Persian with the stipulation that 5 per cent was to be set aside for Gulbenkian. That 5 per cent was made up of equal contributions from Anglo-Persian and Anglo-Saxon.[49]

The outbreak of World War I brought all operations of the TPC to a halt. The British Alien Property Custodian took over the interests of the Deutsche Bank. After the war the British and French signed an agreement in San Remo transferring the Deutsche Bank's share to the French government. The agreement also permitted the government of Mesopotamia (now known as Iraq), over which the British were given a mandate, to receive a 20 per cent share in the company.[50] No U.S. company was given any share in TPC.

Around the same time seven American companies[51] wanted to send geological expeditions into the country. The British, as the mandatory power, refused to grant them permission to enter. At this point the U.S. State Department entered the picture. A note delivered by the U.S. ambassador to Lord Curzon, the British foreign minister, complained that Britain "had given advantage to British oil interests which were not accorded to American Companies and further that Great Britain had been preparing quietly for exclusive control of the oil resources in this region."[52]

The note demanded that "as a general principle any Alien territory which should be acquired pursuant to the Treaties of Peace with the Central Powers must be held and governed in such a way as to assure equal treatment in law and in fact to the commerce of all nations."[53] What the U.S. was in effect demanding was that it, as an Allied power, be treated at least on an equal basis as the French. The problem was that the U.S., unlike Britain and France, never declared war on Turkey and never was a signatory to the peace treaties concluded with Turkey.[54] Furthermore, the mandate system by which Mesopotamia was governed was under the control of the League of Nations, and the United States refused to join the League. The legal bases for its claims in this region were therefore rather tenuous.

In answer to these objections, U.S. Secretary of State Colby replied:

> While it is true that the United States of America was not at war with Turkey, yet it was at war with the principal allies of that country and contributed to the defeat of those allies and, therefore, to the defeat of the Turkish Government.... The United States, as a participant in that conflict and as a contributor to its successful issue, cannot consider any of the associated powers, the smallest not less than itself, debarred from the discussion of any of its consequences, or from participation in the rights and privileges secured under the mandates provided for in the treaties of peace.[55]

In another note the U.S. reiterated the "open door" doctrine, which, in this case, meant equality of treatment for American interests. The U.S. demanded:

> That no exclusive economic concessions covering the whole of any Mandated region or sufficiently large to be virtually exclusive shall be granted and that no monopolistic concessions relating to any commodity or to any economic privilege subsidiary and essential to the production, development, or exploitation of such commodity shall be granted.[56]

It is interesting to note, in this respect, that in its communications with Mexico the U.S. never expressed any concern about the monopolistic practices of its companies in that country. The U.S. note further stipulated that

> . . . in general regulations or legislation regarding the granting of concessions relating to exploring or exploiting economic resources or regarding other privileges in connection with these shall not have the effect of placing American citizens or companies or those of other nations or controlled by American citizens or nationals of other countries at a disadvantage compared with the nationals or companies of the Mandate nation or companies controlled by nationals of the Mandate nation or others.[57]

There was no mention here of the "minimum duty" to the original concession holders, a point which was the main concern of the United States in its negotiations with Mexico. In any event the U.S. finally reached an agreement with Great Britain, which allowed these two countries, together with Holland and France, to have access to Iraq's petroleum resources to the exclusion of everyone else.

The U.S. was able to gain entry for its companies because it controlled most of the world's oil supply. It was in a position to withhold these supplies from recalcitrant nations.

> The American oil industry produced in 1921 about 65 per cent of the world's oil supplies, and purchased about 17 per cent of the remainder (mostly from Mexico). After satisfying all American needs, the industry exported about 58 per cent of total foreign requirements. Anglo-American Oil Co., a former subsidiary of Standard Oil Co. (New Jersey), controlled over 50 per cent of the total business of the United Kingdom. Thus, the American oil companies were in a strong bargaining position. It is not known if they actually used their bargaining strength to the extent of threatening to withdraw from the export market, but the possibility was discussed by officials of Standard Oil Co. (New Jersey).[58]

It should be noted that no one need explicitly mention such a threat. As long as both sides of the bargaining table know that the potential of withdrawal exists, they will adjust their strategies accordingly. Thus, the control of oil carried with it not merely the ability to wage war, which so concerned Churchill and Requa, but also the potential to impose one's will on other nations. As we shall see later, this potential was explicitly recognized by the U.S. in the post–World War II period.

The final agreement over TPC, reached in 1928, granted to the Near East Development Company (NEDC), a corporation established to handle the American interests, a 23.75 per cent share of TPC. This share was equal to that held by each of the other major partners, Anglo-Persian, CFP, and Royal Dutch Shell. The remaining 5 per cent was allocated to Gulbenkian. Standard Oil (New Jersey) and Standard Oil (New York) each received a 25 per cent interest in NEDC. Three other companies—Gulf, Atlantic, and Pan-American Petroleum and Transport, a subsidiary of Standard Oil (Indiana)—each obtained 16⅔ per cent.[59] Subsequently, each of these three sold their shares to the Standard Oil companies so that, by 1934, these two firms held an equal NEDC interest.[60]

These results show how drastically the attitude of the U.S. government had changed towards the Rockefeller interests. Rockefeller and his companies were transformed from pariahs into key instruments of U.S. foreign policy. With the full blessing of the State Department, the negotiations between the seven U.S. oil companies and the TPC were conducted by W. C. Teagle and W. D. Asche, president and vice-president of Standard Oil (New Jersey).[61] Furthermore, by backing the entry of supposedly competing companies into the TPC, the State Department put these companies into a position where they would have to enter into production-sharing arrangements. Such arrangements violated the essence of the U.S. anti-trust laws, the very laws which the U.S. government had used against Standard Oil two decades earlier. In addition, all the partners in TPC had to abide by the "Red Line" agreement, which forbade each one of them to bid individually on concessions in the general area controlled by the old Turkish Empire.[62] All such bids were to be handled through TPC. This stipulation further cartelized the industry and reinforced the tendency for the TPC owners to work together rather than to compete against each other in the market place. Finally, the TPC agreement had the effect of enhancing the power of the Rockefeller-controlled companies by giving them access to prolific reserves of crude. As mentioned in Chapter 3, the Rockefeller companies were traditionally short of crude because of Rockefeller's aversion to the purchase of reserves. The U.S. government, through its actions on the Mesopotamia issue, helped the Rockefeller companies overcome this disadvantage. It also helped the Rockefeller interests obtain reserves in the Netherlands East Indies.

NETHERLANDS EAST INDIES

Before World War I practically all the oil in the Dutch East Indies was controlled by Royal Dutch Shell either directly or through its subsidiaries. The Dutch government ostensibly did not discriminate against non-nationals. Anyone was allowed to apply for an exploration permit and would receive a concession if oil were found. In practice the authorities were reluctant to grant exploration permits to foreigners, thereby giving a virtual monopoly to Shell.

Despite these obstacles, several foreign companies set up subsidiaries in the Indies with the hope of eventually receiving concessions. Among them was a subsidiary of Standard Oil (New Jersey), the Nederland Koloniale Petroleum Maatschappij, which was organized in 1910.[63] During the war the Dutch government amended its mining laws, limiting prospecting licences or concessions to

> (a) Netherlands nationals; (b) inhabitants of the Netherlands or of the Netherlands East Indies; (c) companies domiciled in the Netherlands or in the Netherlands East Indies, of which, in the case of limited companies, the majority of the managers or directors were Netherlands nationals or inhabitants of the Netherlands East Indies.[64]

While the U.S. was somewhat perturbed by this provision, it did not regard it as a serious obstacle to U.S. companies. These companies could, and probably would, require that the managers and directors of their subsidiaries become residents of the Indies. This provision therefore did not seriously conflict with the concept of the "open door." Far more disturbing was another provision, article 28, which severed the link between the discovery of oil and the right to a concession. Under the old mining law, anyone who made a discovery would automatically receive a concession to exploit it. The new law, on the other hand, stated that "only the Government, or companies with which the Government entered into contracts . . . would have the right to extract oil."[65] The U.S. government correctly interpreted this article as a device to exclude non-Dutch companies.

The issue came to a head after the discovery of the Djambi fields in Sumatra. The U.S. consul at Batavia described these fields as "the most valuable mineral oil fields in the whole colony."[66] In 1920 the Dutch government, eager to keep this find under Dutch control, introduced a bill into the States-General granting the entire area to a new company, equally owned by the government and Bataafsche Petroleum Maatschappij, a subsidiary of Shell. This bill thus effectively excluded American interests from the area.

While the bill was being debated, the U.S. expressed its concern to the Dutch government. In the initial discussions the U.S. minister to the Netherlands did not discuss the Djambi bill, but the revisions to the mining law. While admitting that these revisions conformed to the "open door" concept, he objected to them on the grounds of "reciprocity." The State Department complained that "the Netherlands had not accorded the same degree of freedom to American nationals as was accorded to Dutch nationals in the United States" and therefore " had not granted reciprocal privileges to American capital."[67] The Department also noted that, while the Dutch laws apparently did not discriminate, they were administered in such a way as to exclude American firms. Reciprocity, the State Department contended, "must be not only in the letter of the law but in its administration and operation."[68]

In the meantime two American companies—Sinclair and Koloniale, Standard Oil (New Jersey)'s subsidiary—applied for concessions in Djambi. Their applications were denied. The U.S. government protested vigorously. In a note to the Dutch government the American minister stated the essence of U.S. policy:

> . . .the United States has for years carried the burden of supplying a large part of the petroleum consumed by other countries, that Dutch capital has had free access to American oil deposits and that the petroleum resources of no other country have been so heavily drawn upon to meet foreign needs as the petroleum resources of the United States. I have pointed out in the future ample supplies of petroleum have become indispensable to the life and prosperity of my country as a whole, because of the fact that the United States is an industrial nation in which distance renders transportation difficult and agriculture depends largely on labor-saving devices using petroleum products.[69]

This note was a much more explicit formulation of the policy of oil imperialism than that of Requa. The emphasis had, however, somewhat shifted. While Requa had been concerned about the military uses of oil, the American minister to the Netherlands stressed the importance of oil for the economic development of the U.S. He, like Requa, was not content to let the U.S. consumers of oil buy that commodity in the market. The reserves of oil, wherever and in whatever country they existed, must be controlled by U.S. companies. He emphasized this particular point by informing the Dutch that "American capital stands ready to assist in the development of the Djambi fields and other oil deposits in the Netherlands Indies."[70]

The Dutch protested that if the principle of "reciprocity" were universally applied they would have to divide up the Djambi fields among many nationalities. But these protests fell on deaf ears. The U.S. invoked the "non-reciprocity" provisions of its Mineral Leasing Act and refused Shell a permit to explore on federal lands in Utah. Finally, in 1927, the Dutch government introduced a bill to the States-General authorizing the issuance to Koloniale of four tracts of land for exploration and production. The government also promised to continue this policy on the condition that the U.S. declare Holland a "reciprocating country."[71] Such a declaration would again enable Shell, or any other Dutch company, to explore on U.S. federal lands.

Several days later the U.S. government informed the Netherlands that it had accorded Holland recognition "as a reciprocating State within the meaning of the provisions of the Mineral Leasing Act."[72] Thus, by using the hallowed principle of "reciprocity," the U.S. government forced the Dutch government to open its door to Rockefeller's Standard Oil (New Jersey).

As an aside it might be interesting to note how seriously the U.S. viewed "reciprocity." At the same time the State Department was negotiating with the Dutch, the U.S. navy was formulating policy for the use of its oil reserves. The navy wanted to insert a clause, similar to that in the Mineral Leasing Act, restricting concessions to nationals from "reciprocating" countries. The State Department cautioned against the insertion of such a clause, arguing that "it could hardly be of any great practical importance" since the "naval petroleum reserves are less than 1 1/2 per cent of our total petroleum reserves, and the larger part of the naval reserves has already been leased under long term contracts," affording only a very limited opportunity "for foreign-controlled companies to participate in the further development of these reserves."[73] This was a rather revealing statement from an agency which, at the same time, was demanding that the Dutch observe "reciprocity" in "administration and operation" as well as in "the letter of the law."

The State Department had more basic reasons to oppose the insertion of a "reciprocity" clause. First of all, it feared that such a clause "might tend to increase agitation in this country against the operation of foreign petroleum interests in the United States," thus tending "to make more difficult the support by this Department of American petroleum interests abroad." In addition the department was afraid that a "reciprocity" clause would induce other nations also to establish naval reserves and use this clause to exclude U.S. interests. The department was particularly worried about the effects on Brazil, which was "considering a proposal which would declare all petroleum deposits as well as deposits of many of the most important minerals to be reserved for national security and defense and excluding foreign interests from opportunity to participate in their development."[74] "Reciprocity," like "minimum duty" and "open door," hence turned out to be a disposable weapon in the U.S. diplomatic armory. They were to be used when needed and then discarded when they stood in the way of attaining the objective of assuring American control over the world's oil resources.

The U.S. government promoted the expansion of the American oil industry in many parts of the world. The interwar period witnessed a spectacular growth in the overseas operations of U.S. oil companies, especially in Latin America and in the Middle East. Sometimes this expansion was aided by direct State Department pressure. At other times such pressure was not necessary because the leaders of the host country welcomed the investment inflow. But in all countries the leaders knew that behind the oil companies lay the full power of the United States government.

### VENEZUELA

In Latin America, Venezuela was the most important country for the U.S. oil industry. Large-scale investments started flowing there after the drilling by Shell of the Barroso No. 2 well in the Lake Maracaibo area in 1922. The well came in so strongly that it "blew out." This blowout, which spewed forth 850,000 barrels of oil, lasted nine days. The news of this find electrified the oil world. Within days oil men from all over the world were scrambling to obtain concessions in the Lake Maracaibo area.[75]

For most the trip to Venezuela was in vain. They could not obtain concessions because Shell held a virtual monopoly of the on-shore fields. The only concessions available were those on the floor of the lake. Few companies felt that these concessions were worth bidding on.[76] However, two companies, Gulf and Standard Oil (Indiana), thought that they could drill on the lake floor. Gulf obtained concessions in the area along the shore and Indiana in the deeper areas. Since the pool discovered by Shell extended under the lake, these companies were also able to drill highly prolific wells.[77]

These two concessions were the first major American inroads into Venezuelan oil. Up until 1923, when the Americans entered in force,[78] the main companies in Venezuela were Shell and British Controlled Oilfield, a company controlled by the British government.[79] This control had to be kept secret because Venezuelan law forbade the granting of concessions to government-owned companies.[80] Shell obtained its foothold in 1911 through the acquisition of General Asphalt, an American company that had obtained a concession to the area lying east of Lake Maracaibo.[81] British Controlled Oilfield received its concession in 1917.[82]

The entrance of Shell into Venezuela initially did not disturb the Rockefeller oil interests. Venezuela was regarded as a source of asphalt rather than of crude. Furthermore Standard's agents felt that the country was too unhealthy a place.[83] This attitude changed, however, when it became apparent that the country had a large oil potential.

One of the problems Standard Oil (New Jersey) and the other entrants had to face was how to break the British monopoly on concessions. The British had obtained their holdings under the regime of President Juan Vincente Gomez, who had seized power in 1908 with the help of the U.S. navy.[84] Gomez had the habit of granting concessions of mineral resources, which by Venezuelan law belonged to the nation, to his political favourites. They, in turn, resold their concessions to the oil companies. Apparently Gomez himself also exacted a charge for these concessions, and when he died in 1935, he left an estate valued at $200 million.[85] In any event Shell, ever anxious to find new reserves, had taken advantage of this opportunity to obtain concessions at a very low price. That Shell benefited enormously from this system is evident from the praise Sir Henri Deterding, the head of Shell, lavished on Gomez shortly before the latter's death:

> I satisfied myself. . .that the government under General Gomez appeared sound and constructive and likely to be fair to foreign vested interests. And now that I know Venezuela better, I can testify that in his 26 years of virtual dictatorship, General Gomez has consistently insisted on fair play to foreign capital.[86]

In order to break the British monopoly, the U.S. State Department persuaded the Venezuelan government to force Shell to surrender the concessions it had not explored.[87] These lands were, in turn, sold to Gomez's supporters, who resold them to the oil companies. The State Department also informed the Venezuelan government that a British company, presumably British Controlled Oilfield, was controlled by the British government and, therefore, not entitled to concessions.[88] On State Department orders, the U.S. minister arranged a private meeting between Gomez and representatives of Sinclair Oil.[89] Most important of all was the role played by the U.S. government in revising Venezuela's basic petroleum law. In 1921, the U.S. minister cooperated with Standard Oil (New Jersey), Sun Oil, and other companies in drawing up

**Table 4.3: Major American Companies Entering Venezuela after Passage of 1922 Petroleum Law**

| *Company* | *Controlling Interest* | *Year of Entry* |
|---|---|---|
| Orinoco Oil | Pure Oil | 1923 |
| Venezuelan Gulf | Gulf | 1923 |
| Lago Petroleum | Doheny[a] | 1923 |
| Richmond Petroleum | Standard Oil of California | 1925 |
| Venezuela Atlantic Refining | Atlantic Refining | 1926 |
| California Petroleum | Texas Company | 1927 |

a. Standard of Indiana purchased Lago from Doheny in 1925.

Sources: L. Vallenilla, *Oil: The Making of a New Economic Order, Venezuelan Oil and OPEC* (McGraw-Hill, New York, 1975), p. 37, and Luis Alcala Sucre, "The Impact of Multinational Oil Companies in Venezuela: The Oil Companies in General and Gulf Oil in Particular" in Jon P. Gunneman (ed.), *The Nation-State and Transnational Corporations in Conflict with Special Reference to Latin America* (Praeger, New York, Washington and London, 1975), p. 184.

proposals to change the law in order to "make it possible for foreign oil companies to extend their oil operations."[90] Their proposals were incorporated in the Petroleum Law of 1922, which the U.S. oil companies regarded as the best in Latin America.[91] Following the enactment of this law, a number of important companies set up operations in Venezuela. They are listed in Table 4.3.

Standard Oil (New Jersey) and Sun were the only important U.S. companies with holdings in Venezuela before the enactment of the law. Standard entered the country in 1921 through a subsidiary, Standard Oil Company of Venezuela.[92] It was able to pick up its early concessions only by learning the "rules of the game" from Sun. The company approached Gomez with a bid on five concessions. Gomez's son-in-law also bid on those concessions, obtained them, and promptly resold them to Sun. From that day on, Standard dealt directly with Gomez's son-in-law.[93]

The turmoil in Mexico also played a role in the movement of many U.S. firms to Venezuela. The persistent agitation for the nationalization of sub-surface mineral rights caused many firms to divert investment from Mexico to Venezuela. Furthermore, the companies in Mexico pumped the fields as quickly as they could, thereby rapidly depleting the nation's oil. Production, after peaking in 1921, started declining. The discovery of prolific oil areas in Venezuela offered a welcome substitute to Mexico. Thus the rise of oil in Venezuela was accompanied by the decline of oil in Mexico.

The Mexican events were an important factor causing Doheny to move to Venezuela. His Venezuelan subsidiary, Lago Petroleum, obtained prize concessions on the floor of Lake Maracaibo. In 1925 he sold his holdings to Standard Oil (Indiana),[94] making Indiana one of the major companies in the country. By 1929, a year after Venezuela had become the second largest producer of oil in the world,[95] three companies—Shell, Gulf, and Indiana—dominated the scene. Shell had 45 per cent of production, the other two, 27 per cent each. Venezuelan oil provided one-half the production of Shell and Gulf, and more than one-half that of Indiana.[96]

The developments in Mexico also affected the location of refineries. Despite Gomez's generosity to the oil companies, the companies feared that the same nationalist sentiments which swept Mexico could also sweep Venezuela. As a precaution, Shell decided to build a giant refinery on the Dutch-owned Caribbean island of Curacao. By doing so, it initiated a policy that was to be followed by all the other international oil companies: that of locating refineries outside the political jurisdiction of the host countries. Since crude oil is worthless unless it is refined, the host countries cannot sell their crude if they have no refineries. Thus the locational policy of the oil companies greatly enhanced their bargaining power *vis-à-vis* the host countries. If the latter either increased their royalty demands or nationalized the fields, the companies would merely refuse to refine the crude.

Standard Oil (Indiana) followed Shell by constructing a major refinery on another Dutch Caribbean island, Aruba. The Dutch apparently were reluctant to grant permission for this refinery, which would compete with Shell. However, they were then involved in the dispute with the U.S. over Djambi and did not want to antagonize the U.S. further by denying Standard of Indiana permission for the refinery.[97] This refinery, whose main market was the U.S., began operations in 1929.[98] The main market of the Shell refinery was Europe.[99] These two refineries processed about one-third of Venezuela's crude.[100]

Indiana's refinery on Aruba opened at a particularly unpropitious time, the eve of the Great Depression. The reduction in demand brought downward pressure on prices. Accompanying this decline in demand was a significant increase in supply, following major discoveries in Oklahoma and in East Texas. This combination of decreased demand and increased supply severely depressed prices in the States. Added to this was the downward pressure of increased imports from Venezuela. Soon a call arose for the restriction of imports. The U.S. government in 1933, as part of the National Industrial Recovery Act program, imposed controls on imports, limiting them to "the daily average of petroleum and petroleum products during the last six months of 1932."[101] These controls would have struck a virtual death blow to Standard Oil (Indiana)'s foreign operations; since Indiana had no marketing facilities outside the United States, it could not divert its foreign production to other markets. In anticipation of these controls, it had sold its Venezuela properties, including the refinery on Aruba, to Standard Oil (New Jersey) in 1932.[102] Jersey then exported refined products from Aruba to Europe, thereby complying with the U.S. import program. At the same time it reduced its refined product exports from the U.S.[103]

Following this acquisition, Jersey, which in 1929 accounted for less than one per cent of Venezuela's production, became by 1935 the largest company in Venezuela, producing more than one-half of total output.[104] By the end of the 1930s, Jersey, Shell, and Gulf controlled the bulk of Venezuela's production. Venezuela was the most important oil producer outside of the United States. Its importance was explicitly recognized by Nelson Rockefeller, who in 1935 deliberately put himself on the board of Jersey's subsidiary, Creole.[105] He regarded Creole as the plum of the Rockefeller empire since it was the purchase of the Venezuelan properties that made Jersey the largest oil company in the world.[106] Despite its early lack of success in Venezuela, Jersey eventually became the kingpin of the Venezuelan economy. In this it was aided, at least indirectly, by U.S. government policy.

### CANADA

In Canada, U.S. oil interests were able to enter the scene without the assistance of the U.S. State Department. Imperial Oil, the subsidiary of Rockefeller's Standard Oil (New Jersey), managed to obtain control of Canada's main production centre, the Turner Valley, just outside of Calgary. Oil had been discovered in the Turner Valley in 1914 by the Calgary Petroleum Products Company.[107] Among those associated with this company were Robert Bennett, later to become prime minister of Canada,[108] and Senator James Lougheed,[109] the grandfather of Alberta's Premier Peter Lougheed.

Despite its initial success, the company ran into financial difficulties. In 1920 fire destroyed its absorption plant. Lacking funds, the company turned to Imperial Oil for help. Imperial responded by granting the funds in return for a controlling interest. Upon obtaining control, Imperial reorganized the company and renamed it Royalite.[110] From this toehold, Imperial and Royalite, in a manner reminiscent of the Rockefeller of early days, fanned out to seize control of the entire Turner Valley. This seizure had serious economic consequences for Alberta, as explained by Richards and Pratt:

> . . . Royalite and other Imperial-owned companies consolidated their control over Turner Valley, provoking much animosity among the many small independents in the field. Royalite had exclusive marketing arrangements with Calgary's gas utility and it also owned the only refineries purchasing naphtha and other oil products from the field. The independents had no markets for gas, so they simply flared it off; in the process they thereby reduced the energy drives in the field itself. . . it was clear that, failing drastic political intervention, Turner Valley would quickly be depleted.[111]

Gray estimates that by 1932 gas was flared at a rate of up to 600 million cubic feet per day.[112] Over the life of the Turner Valley field, approximately 1.8 trillion cubic feet of gas was flared.[113] Since natural gas drives oil to the surface, and since flaring reduces its pressure, flaring of natural gas reduces the amount of oil that can eventually be recovered. According to Gray, engineers have estimated that, over its life, the Turner Valley field produced 120 million barrels of oil. If flaring had not occurred, more than 500 million barrels might have been produced. The "value of the oil reserves which were lost and the gas which was wasted amounted to more than a billion dollars," wrote Gray in 1970.[114] Had that billion dollars been invested in 1970 at a modest rate of 6 per cent, it would have totalled over two billion dollars in 1983—equal to $1,000 for each man, woman and child in Alberta. This is a loss which, in real terms, can never be recovered.

During World War II, Alberta's Oil and Gas Conservation Board became concerned about the extent of flaring. Based on the advice of American conservation engineers, a plan to reduce gas wastage was worked out. Imperial, acting in a manner similar to the way the oil companies had behaved in Mexico (see page 63), refused to comply with this plan, stating that "the economics of the situation were against it." Imperial, instead, demanded $500,000 from Ottawa as its price for implementing the plan.[115] This demand for concessions from governments in return for carrying out public policy became an integral part of Imperial's behaviour pattern in the post–World War II years.

Imperial was unsuccessful in obtaining this grant. Under pressure from Alberta and the independent producers, it agreed to a compromise under which it gave up its exclusive arrangement with Calgary's gas company. In return the independents would use Imperial's pipeline facilities.[116] This control of pipeline facilities gave Imperial effective control over production by the independents in the Turner Valley.

In addition to expanding into the Turner Valley, Imperial also moved into the Northwest Territories. According to Richards and Pratt, Imperial entered the Territories partly in response to fears that Royal Dutch Shell (known among Standard personnel as "the yellow peril") was moving into the Canadian Northwest.[117] Imperial discovered oil near Norman Wells in 1920,[118] and built a small refinery at Norman Wells a year later. After operating about a year, the refinery was shut down. It reopened in 1933 to supply fuel to the Great Bear Lake mining divisions. In 1939 Imperial built a modern refinery at Norman Wells with a capacity of 840 barrels per day. It was used mainly to refine aviation gasoline.[119]

### MEXICO

As in Canada and Venezuela, petroleum investment in the rest of the Western Hemisphere significantly expanded during the interwar period. U.S. firms entered a number of Latin American countries, among them Peru, Bolivia, and Colombia. This expansion was not continuous, however. In two countries, Bolivia and Mexico, investment fell because of expropriations. Bolivia expropriated the properties of Standard Oil (New Jersey) in 1937;[120] Mexico nationalized its oil in 1938.[121]

The Mexican nationalization occurred because the oil companies, by refusing to obey an order of the Mexican Supreme Court, forced the government to act against them. The Mexican government at that time had no policy for nationalization. It was content to let the companies operate in peace. It was, however, forced to intervene in a strike against the companies. It appointed a number of bodies to investigate the workers' grievances, and these bodies upheld the workers' demands, especially those for wage increases, on the grounds that the companies were extremely profitable and could well afford to pay their workers. The companies appealed these rulings to the Mexican Supreme Court, which sustained the findings. The companies refused to obey the court because, they said, they could not afford to do so. Following this refusal the government expropriated the properties of Shell, Standard Oil (New Jersey), Sinclair, and Standard Oil of California.[122]

The companies retaliated by imposing a boycott on Mexico. They threatened all companies buying Mexican oil with lawsuits, including those companies selling machinery to the Mexican oil industry. They refused to sell Mexico any materials, and the Mexican government could not obtain replacement parts.[123] They also asked their home governments to exert diplomatic pressure on Mexico. While both the United States and Britain protested vigorously, they did not go beyond making these formal protests.

There were a number of reasons for this moderation in U.S. policy. First, the U.S. ambassador to Mexico, Josephus Daniels, was sympathetic towards the Mexican government. He blamed the expropriation entirely on the ineptness of the companies. In a letter to Secretary of State Cordell Hull, who wanted to take a more vigorous approach towards Mexico, Daniels wrote: "I wonder at the ineptness of the representatives of the oil companies.... Some are so dumb that if they had to start a business on their own, it would be foredoomed to failure. Initiative and tact are not in their vocabulary..."[124] Daniels's arguments were convincing enough to dissuade Hull from intervening militarily.

Second, Mexico was no longer important in the world oil picture. Its production had fallen from its peak of 193 million barrels in 1921 to 47 million barrels in 1937. Of this latter amount only 54 per cent was exported, while in the heyday of Mexican production over 80 per cent was exported.[125] Thus the nationalization did not have a major impact on the world oil industry.

Third, the U.S. government under President Franklin Roosevelt was anxious to improve its relations with Latin America. Officially the "Good Neighbor Policy" displaced "dollar diplomacy." Intervention in Mexico would have destroyed this new policy. Furthermore, it was feared that it might spur demands for nationalization in Venezuela, which was far more important to both the oil industry and the U.S. government. Intervention in Mexico was not worth such a risk.

When it became apparent to the oil companies that the U.S. government was not going to intervene, they began to change their tune. Initially they refused to consider any offer by Mexico for compensation, demanding instead the return of their properties. Finally, in 1940, Sinclair accepted an offer of $8.5 million. Others followed suit. All told the Mexican government paid out $165 million in principal and interest.[126]

The Mexican episode illustrates the importance to the oil companies of backing by their home governments. With such backing they can treat the host governments as colonial lackeys, whom they can order around at will. Without such backing they have to accede to the demands of the host governments.

## MIDDLE EAST

While in terms of production and investment Latin America was the most important area of expansion for U.S. oil companies in the interwar period, the Middle East also attracted their attention. This area was destined to replace Latin America as the key oil area in the post–World War II era. As has already been discussed, the first entry of U.S. firms in the Middle East was in the TPC, which changed its name in 1929 to IPC (Iraq Petroleum Company).[127]

Around the same time as the U.S. was negotiating to allow American companies in Iraq, the Gulf Oil Company obtained an option to purchase concessions in Bahrein and Kuwait from Eastern and General Syndicate, a British company. This company, headed by Major Frank Holmes, a New Zealander, had obtained these concessions in 1925. When Gulf was ready to exercise its options, it had already joined IPC and was bound by its "Red Line" agreement, which prohibited IPC owners from bidding on their own for concessions in most of the territory of the old Ottoman Empire. Bahrein was within the "Red Line" but Kuwait was on the outside.

Gulf had to ask the IPC for permission to exercise its option in Bahrein. The IPC, fearing the entry of another American firm in the area, refused. Gulf then sold its option to Standard Oil of California, which was not a participant in IPC and therefore not bound by the "Red Line" agreement.[128] The British, who exercised control over Bahrein, tried to prevent Standard Oil of California from obtaining the concession. The U.S. State Department intervened and the concession was approved in 1930.[129]

Since Kuwait was outside the "Red Line," Gulf did not need approval from IPC to exercise its option there; however, it was blocked by the British government from doing so.[130] The British insisted that the concession be given to a British company. The State Department protested, demanding U.S. companies be given equal treatment. At about the same time Anglo-Persian, a partner in IPC, put in a bid for a concession. After much haggling, Gulf and Anglo-Persian decided to submit one bid. In 1933 they formed the Kuwait Oil Company, in which each held an equal share, to operate their Kuwait concession.[131]

Saudi Arabia was the next important area of American penetration in the Middle East. As in the rest of the Middle East, the Americans were not the first to enter. Major Holmes's Eastern and General Syndicate obtained a concession from King Ibn Saud in 1923.[132] Though the U.S. government had no direct interest in that area and had no diplomatic or consular officers in the country, it nevertheless instructed its diplomats elsewhere in the Middle East to obtain the details of the concession. The American vice-consul in Bushire, Persia, reported back the details which, he claimed, were furnished him by "a British subject, the agent of Standard Oil." He warned that this "source in particular should be kept confidential."[133] The vice-consul also noted that "the concession is reported to be with the consent of the Colonial Office."[134] He then commented that:

> It is generally agreed that there is no real competition between the Eastern & General Syndicate, and the Anglo-Persian Oil Company. A. T. Wilson [General Manager of Anglo-Persian] tried to get the same concession for the latter company, and also tried to get a concession in Koweit [sic]. His failure seems not to worry anyone, and may have been more or less deliberate.[135]

Sir Arnold Wilson apparently believed that there was no oil in Saudi Arabia. So also did King Ibn Saud. He accepted advances from Eastern and General never believing that oil would be found.[136]

Saud's judgement was correct as far as Eastern and General was concerned. Holmes's company found no oil and let the concession lapse. The increasing demands of Ibn Saud left Holmes no other choice. A 1926 dispatch from the American consulate in Baghdad indicates the nature of some of these demands:

> I learn that Major Frank Holmes considers that his oil concession. . . has lapsed. It is said that, in order to get this concession in the first place, Major Holmes was obliged to pay Ibn Saud a considerable sum of money in cash and also to give him an expensive motor car. . . now Ibn Saud is demanding more and further presents from Major Holmes if the concession is to remain in force. I am informed that apparently the money and presents are not forthcoming and it was when speaking about the new demands of Ibn Saud that Major Holmes remarked that his concession had lapsed.[137]

After several years Standard Oil of California started negotiating with Ibn Saud for the lapsed concession. Though the State Department did not directly participate in the negotiations, it kept close tabs on developments. State Department participation was not necessary because King Ibn Saud, unlike most of the other Middle East potentates, was not under British control. The British backed his rivals, the Hashemites, in the battle with Ibn Saud's Wahabis in the post–World War I struggle for control of the country. Hence Saud was very receptive to American investments as a counterweight to British influence in the Middle East.

An important negotiator with King Ibn Saud was Kenneth Twitchell, an independent American geologist, who acted on behalf of Standard Oil of California. He also kept the State Department informed about the negotiations. A State Department memorandum of 1932 summarized Mr. Twitchell's impressions as follows:

> Mr. Twitchell said that the present was an excellent time to enter the scene in Saudi Arabia. King Ibn Saud, he asserted, was disposed to be very friendly to the United States and was hopeful that Americans, who had no political ax to grind, would become interested in the exploitation of the natural resources of his Kingdom. The pilgrimage business, from which Saudi Arabia derives most of its revenue, has been seriously hit by the depression, and the revenue that might be expected from oil or other mineral development would be especially welcome at this time.[138]

As an indication of the struggle for control over Middle East oil resources, Twitchell also discussed the attitudes of the foreigners in Saudi Arabia towards the United States. The State Department summary states that Twitchell described the Dutch as "the most friendly" while the British "would be sure to look with suspicion and disfavor on the entrance of American interests."[139]

Recognizing the British antagonism to American entry into the area, the State Department tried to adopt a position that was halfway between giving into British desires to exclude American influence, on the one hand, and ousting Britain as the dominant power in the region, on the other. This ambivalent position was evident in a reply to a telegram sent to the State Department in 1932 by Francis B. Loomis, representative of Standard Oil of California. After informing the department that his company was about to sign a contract with King Ibn Saud, he asked, "What protection could we expect in case of disorders in Arabia or in case the present government is overthrown and its successor were disposed to break our contract without just cause?"[140] The department's reply was ambiguous. It stated:

> The Department is unable to indicate in advance the nature of the protection, if any, which it could accord in the event of the contingencies to which you refer, but you will doubtless appreciate the possible difficulty of assuring effective protection for American influence.[141]

The "difficulty" presumably would arise from the dispatch of a military force to an area regarded as in the British sphere of influence. The Middle East, after all, was not Latin America.

Despite the lack of a firm guarantee of protection from the U.S. government, the company went ahead with its negotiations. In May, 1933 it signed an agreement with the government of Saudi Arabia. This agreement, in addition to stipulating the area and time period of the concession, the terms and methods of payment, and many technical details, provided that "the enterprise . . . shall be directed and supervised by Americans."[142] In other words, this Saudi oil was to be considered as American oil.

Apparently the terms of the agreement were highly favourable to the company. For instance, they exempted the company from paying duties on its imports and gave the company the right to use, free of charge, the government's water and other natural resources.[143] Though the agreement called for publication "in Saudi Arabia in the usual manner,"[144] it was not published in its entirety. In a letter to the State Department, Twitchell noted that the "Government is not publishing details of the loans and payments, as it wishes to avoid the criticism of those who might say it should have obtained terms equal to the Iraq and Persian governments."[145]

In any event the company was not able to enjoy any profits from its investment until 1938, when it discovered oil in commercial quantities.[146] In the meantime it brought in a partner, the Texas Company, in 1936. They created the Arabian-American Oil Company (ARAMCO), a company in which each held a 50 per cent share. Standard Oil of California brought in Texas because it needed marketing facilities for its production from Bahrein, where it had discovered oil in 1932. It was reluctant to sell its oil in the States, where it had marketing facilities, because its Bahrein oil would compete with its California oil. Since it had no marketing facilities outside the States, it tried to make an arrangement with the Big Three—Shell, Anglo-Persian, and Standard Oil (New Jersey), who controlled most of the world's marketing facilities. But no agreement could be reached. It then turned to Texas, which had extensive marketing facilities in Europe, Australia, Africa, China, and other areas of the Far East. Texas was supplying these areas from the United States, but found it more profitable to supply them from the Middle East. Therefore it had an incentive to enter into arrangements with Standard Oil of California.[147]

With the entry of American companies into Saudi Arabia, the U.S. had an interest in every important producing area of the Middle East with the exception of Iran. That country was completely controlled by the British, who were successful in excluding everyone else during the interwar period. For many years Persia was the only producer of oil in the Middle East, giving the British a virtual monopoly of Middle East production throughout the 1920s. During the 1930s production com-

**Table 4.4: Per Cent Control by Major Powers in Key Producing Areas Outside the United States, 1929 and 1939**

| *Power* | *Producing Area* | | |
|---|---|---|---|
| | *Western Hemisphere* | *Near East* | *Far East* |
| 1929 | | | |
| United States | 58 | – | 5 |
| Britain and Holland | 39 | 100 | 90 |
| Other | 3 | – | 5 |
| 1939 | | | |
| United States | 48 | 16 | 19 |
| Britain and Holland | 36 | 78 | 73 |
| Other | 16 | 6 | 8 |

Source: H. Williamson, R. Andreano, A. Daum & G. Klose, *The American Petroleum Industry, The Age of Energy, 1899–1959* (Northwestern University Press, Evanston, Ill., 1963), p. 730.

menced in Iraq, Bahrein, and Saudi Arabia. As a result the British share of Middle East production fell and the non-British, especially the American, share rose.

A similar pattern prevailed in the Far East. In this area the dominance of the British-Dutch firms was eroded as the American share rose. Only in the Western Hemisphere did the American share fall, due mainly to the nationalizations in Bolivia and Mexico and the insistence by some Latin American countries that their own nationals be permitted to participate in oil operations. The fall in the U.S. share was, moreover, not accompanied by a rise in the British share but rather by a rise in the share held by Latin American nationals. To a certain extent this meant a rise in the U.S. share since some, but by no means all, Latin American shareholders were subject to U.S. influence and control. These control patterns are summarized in Table 4.4.

These changes in control patterns followed the changes in investment patterns. As was already shown in Table 4.1, practically all investment in production in 1919 was in the Western Hemisphere, with more than three-fifths concentrated in Mexico. By 1935, the latest year for which data were compiled on a basis comparable with earlier years, there were two important trends evident in the geographical distribution of investment in production. First, there was a distinct shift from the Western to the Eastern Hemisphere, with the former's share declining from 98 to 85 per cent. Had comparable data been compiled for 1939,

**Table 4.5: Geographic Distribution of U.S. Direct Foreign Investment in Oil Production,[a] 1919 and 1935**

| Country | *1919* Investment ($ Millions) | % | *1935* Investment ($ Millions) | % |
|---|---|---|---|---|
| Western Hemisphere | 321 | 98 | 742.5 | 85 |
| Venezuela | (18) | ( 5) | (240) | (28) |
| Mexico | (200) | (61) | (206) | (24) |
| Colombia | (20) | ( 6) | (126) | (14) |
| Peru | (45) | (14) | ( 60) | ( 7) |
| Canada | (30) | ( 9) | ( 55) | ( 6) |
| Aruba | (—) | (—) | ( 45) | ( 5) |
| Trinidad | ( 5) | ( 2) | ( 7) | ( 1) |
| Central America | ( 3) | ( 1) | ( 3.5) | ( b) |
| Eastern Hemisphere | 7.5 | 2 | 130 | 15 |
| Netherlands East Indies | (—) | (—) | ( 75) | ( 9) |
| Iraq, Palestine, Syria and Cyprus | (—) | (—) | ( 25) | ( 3) |
| Rumania | ( 7.5) | ( 2) | ( 20) | ( 2) |
| Arabia, including Bahrein | (—) | (—) | ( 10) | ( 1) |
| Total | 328.5 | 100 | 872.5 | 100 |

a. Includes refining.
b. Less than one-half of one per cent.

Source: C. Lewis, *America's Stake in International Investment* (Brookings Institution, Washington, 1938), p. 538.

this shift would have been even more pronounced, as a result of the nationalizations in Mexico and Bolivia. Within the Eastern Hemisphere, most of the investment was concentrated in the Netherlands East Indies. The Middle East had not yet come into full bloom.

The second important trend, within the Western Hemisphere, was a marked shift away from Mexico towards Venezuela, which became the prime outlet for U.S. investment. Mexico's share, even before nationalization, had fallen sharply, from more than three-fifths to less than one-quarter. These trends are shown in Table 4.5.

During the interwar period the pattern of petroleum investment changed on a functional as well as on a geographic basis. As was mentioned in Chapter 3, the pre–World War I investments were essentially market-oriented. Following the war, they became supply-oriented, stemming from (1) the policy shift of the Rockefeller group from one of purchasing crude in the open market to one of outright acquisition of reserves, and (2) the adoption by the State Department of a policy

**Table 4.6: Functional Distribution of U.S. Direct Foreign Investment in Oil, 1914, 1919, and 1935**

| | *Total* | *Market-Oriented* (Distribution) | *Supply-Oriented* (Production)[a] |
|---|---|---|---|
| *1914* | | | |
| Investment | | | |
| ($ Millions) | 343 | (200) | (143) |
| (%) | 100 | ( 58) | ( 42) |
| *1919* | | | |
| Investment | | | |
| ($ Millions) | 604 | (275) | (329) |
| (%) | 100 | ( 45) | ( 55) |
| *1935* | | | |
| Investment | | | |
| ($ Millions) | 1,381.5 | (509) | (872.5) |
| (%) | 100 | ( 37) | ( 63) |

a. Includes refining.

Source: C. Lewis, *America's Stake in International Investment* (The Brookings Institution, Washington, 1938), pp. 579, 588.

**Table 4.7: Per Cent Changes in U.S. Direct Foreign Investment by Industry, 1919–35**

| *Industry* | *Change* |
|---|---|
| Utilities | 688 |
| Manufacturing | 135 |
| Petroleum[a] | 129 |
| Other[b] | 44 |
| Mining and Smelting | 39 |
| Sales Organizations[c] | 34 |
| Agriculture and Timber | 0 |
| Railroads | –12 |
| Total | 86 |

a. Includes exploration, production, refining, and distribution.
b. Mainly banks and insurance companies.
c. Includes trading companies and sales branches and subsidiaries of large corporations. Excludes petroleum distribution.

Source: C. Lewis, *America's Stake in International Investment* (The Brookings Institution, Washington, 1938), p. 605.

encouraging U.S. nationals to obtain reserves abroad. As can be seen in Table 4.6, almost three-fifths of U.S. investments in petroleum in 1914 was in distribution, *i.e.*, market-oriented. By 1935, the distribution share dropped to less than two-fifths. The control of reserves became the crucial element in controlling both the international industry and the economies of other countries.

Along with the growing importance of production within total oil investments, oil became an increasingly important part of all investments. Between the years 1919 and 1935, as Table 4.7 shows, oil investments grew at a faster pace than total investments. This contrasts with the period between 1897 and 1914, when oil investments grew at a slower pace than total investments. Furthermore, in the pre–World War I period, oil ranked fifth in rate of growth while in the postwar period it ranked third.

This increase in the relative importance of oil industry investment stemmed directly from the U.S. government policy of affording special treatment to oil investments abroad. In contrast to the pre–World War I era, when oil was treated as just another commodity, oil became the key commodity in U.S. foreign economic policy. This interest in oil on the part of the government forged strong links between the industry and the State Department, links that were to grow even stronger in the post–World War II years.

## Notes

1. A. Sampson, *The Seven Sisters, The Greatest Oil Companies and The World They Shaped* (The Viking Press, New York, 1976), p. 49.
2. Ibid.
3. H. O'Connor, *World Crisis in Oil* (Monthly Review Press, New York, 1962), p. 280.
4. Ibid.
5. Sampson, *The Seven Sisters*, p. 48.
6. Ibid.
7. Ibid., p. 49.
8. Ibid., p. 47.
9. Ibid., p. 48.
10. O'Connor, *World Crisis in Oil*, p. 281.
11. Sampson, *The Seven Sisters*, p. 52.
12. O'Connor, *World Crisis in Oil*, p. 53.
13. J. Stork, *Middle East Oil and the Energy Crisis* (Monthly Review Press, New York and London, 1975), pp. 8–9.

14. C. Tugendhat & A. Hamilton, *Oil, The Biggest Business* (Eyre Methuen, London, 1975), p. 66.
15. O'Connor, *World Crisis in Oil*, p. 279.
16. Tugendhat & Hamilton, *Oil, The Biggest Business*, p. 66.
17. Great Britain, *5 Parliamentary Debates* (Commons), LV (1913), pp. 1475, 1477.
18. O'Connor, *World Crisis in Oil*, p. 282.
19. Sampson, *The Seven Sisters*, p. 55.
20. Tugendhat & Hamilton, *Oil, The Biggest Business*, p. 68.
21. J. E. Hartshorn, *Oil Companies and Governments, An Account of the International Oil Industry in its Political Environment* (Faber & Faber, London, 1962), p. 233.
22. P. de la Tramerye, *The World Struggle for Oil* (Knopf, New York, 1924), p. 106.
23. Ibid., p. 10.
24. U.S. Federal Trade Commission, *The International Petroleum Cartel* (U.S. Government Printing Office, Washington, 1952), p. 54.
25. U.S. Fuel Administration, *Final Report of the United States Fuel Administration, 1917–19* (Washington, U.S. Government Printing Office, 1921), p. 261.
26. Ibid., p. 277.
27. Ibid., p. 271.
28. Ibid., p. 272.
29. In actuality the British, despite their misgivings about the mixed nationality of Shell, regarded Shell's reserves as part of Britain's reserves. In this instance their policy was exactly the same as that of the Americans. See Hartshorn, *Oil Companies and Governments*, p. 233.
30. *Historical Statistics of the United States, Colonial Times to 1957: A Statistical Abstract Supplement* (U.S. Government Printing Office, Washington, 1960), p. 565.
31. S. Nearing & J. Freeman, *Dollar Diplomacy*, reprint (Monthly Review Press, New York and London, 1966), pp. 85–89.
32. Ibid., pp. 89–90.
33. C. Lewis, *America's Stake in International Investment* (The Brookings Institution, Washington, 1938), p. 222.
34. D. Lawrence, *The True Story of Woodrow Wilson* (Doran, New York, 1924), p. 100.
35. Nearing & Freeman, *Dollar Diplomacy*, pp. 96–97.
36. Ibid., p. 107.
37. Ibid., p. 111.
38. Ibid., pp. 111–12.
39. Lewis, *America's Stake in International Investment*, pp. 221–22.
40. H. Williamson, R. Andreano, A. Daum & G. Klose, *The American Petroleum Industry, The Age of Energy, 1899–1959* (Northwestern University Press, Evanston, Ill., 1963), p. 262.
41. Ibid., p. 29.
42. M. Wilkins, *The Maturing of Multinational Enterprise: American Business Abroad, 1914-1970* (Harvard University Press, Cambridge, 1974), p. 36.

43. In actuality Mexico moved into second place in 1918. This change in relative position was caused not only by the rapid expansion of Mexican production, but also by the decline in Russian production following the Revolution and the Civil War.
44. Williamson, Andreano, Daum & Klose, *The American Petroleum Industry*, p. 29.
45. H. Fraser, *Diplomatic Protection of American Petroleum Interests in Mesopotamia, Netherlands East Indies, and Mexico. A Study Prepared for the Special Committee Investigating Petroleum Resources*. U.S. Senate. 79th Congress. 1st Session. Document No. 43 (U.S. Government Printing Office, Washington, 1945), p. 59.
46. Ibid., pp. 72–73.
47. Williamson, Andreano, Daum & Klose, *The American Petroleum Industry*, p. 731.
48. U.S. Federal Trade Commission, *The International Petroleum Cartel*, pp. 48–49.
49. Ibid., p. 50.
50. Ibid., pp. 50–51.
51. The companies were: Mexican Petroleum, controlled by Doheny; Texas; Gulf; Atlantic; Sinclair; Standard Oil (New York); and Standard Oil (New Jersey). See ibid., p. 52.
52. Fraser, *Diplomatic Protection*, pp. 3–4.
53. Ibid., p. 3.
54. The Allies, with the exception of the United States, signed two peace treaties with Turkey. The first, the Treaty of Sèvres, was signed in 1920 but was never put into effect because of Kemal Ataturk's revolution in Turkey. The second, the Treaty of Lausanne, was signed in 1923. The U.S. signed a treaty of amity and friendship with Turkey in 1923 but this treaty was not ratified by the Senate. See ibid., p. 2.
55. Ibid., p. 10.
56. Ibid., p. 5.
57. Ibid.
58. U.S. Federal Trade Commission, *The International Petroleum Cartel*, p. 53.
59. Ibid., p. 54. Of the seven companies listed in note 51 above, two—Sinclair and Texas—dropped out of the negotiations. The interests of Mexican Petroleum were taken over by Pan American.
60. Ibid. In 1933 Standard Oil (New York) merged with the Vacuum Oil Company and changed its name to the Socony-Vacuum Oil Company.
61. Ibid., p. 53.
62. Ibid., pp. 65–67.
63. Fraser, *Diplomatic Protection*, p. 32.
64. Ibid., p. 33.
65. Ibid.
66. Ibid., p. 34.
67. Ibid., p. 35.
68. Ibid., p. 38.
69. Ibid., p. 41. The note was dated April 12, 1921.

70. Ibid., p. 42.
71. Ibid., pp. 45–46.
72. Ibid., p. 48.
73. Ibid., p. 51.
74. Ibid.
75. Luis Alcala Sucre, "The Impact of Multinational Oil Companies on Venezuela: The Oil Companies in General and Gulf Oil in Particular" in J. P. Gunneman (ed.), *The Nation-State and Transnational Corporations in Conflict with Special Reference to Latin America* (Praeger, New York, Washington and London, 1975), pp. 177–223.
76. Ibid., p. 182.
77. O'Connor, *World Crisis in Oil*, pp. 135–36.
78. Neil Jacoby, *Multinational Oil, A Study in Industrial Dynamics* (Macmillan, New York, 1974), p. 32.
79. Luis Vallenilla, *Oil: The Making of a New Economic Order, Venezuelan Oil and OPEC* (McGraw-Hill, New York, 1975), p. 7.
80. O'Connor, *World Crisis in Oil*, p. 131.
81. Vallenilla, *Oil: The Making of a New Economic Order*, p. 7.
82. Ibid.
83. O'Connor, *World Crisis in Oil*, p. 131.
84. Ibid., pp. 129–30.
85. Tugendhat & Hamilton, *Oil, The Biggest Business*, p. 79.
86. O'Connor, *World Crisis in Oil*, p. 130.
87. Ibid., p. 132.
88. Ibid., p. 131.
89. Ibid.
90. Wilkins, *The Maturing of Multinational Enterprise*, p. 115.
91. Ibid.
92. Vallenilla, *Oil: The Making of a New Economic Order*, p. 36.
93. O'Connor, *World Crisis in Oil*, p. 131.
94. Wilkins, *The Maturing of Multinational Enterprise*, p. 115.
95. O'Connor, *World Crisis in Oil*, p. 136.
96. Ibid.
97. Vallenilla, *Oil: The Making of a New Economic Order*, pp. 30–31.
98. Wilkins, *The Maturing of Multinational Enterprise*, p. 115.
99. Vallenilla, *Oil: The Making of a New Economic Order*, p. 29.
100. Williamson, Andreano, Daum & Klose, *The American Petroleum Industry*, p. 541.
101. E. H. Shaffer, *The Oil Import Program of the United States, An Evaluation* (Praeger, New York, Washington and London, 1968), p. 9.
102. Wilkins, *The Maturing of Multinational Enterprise*, p. 209.
103. Ibid., p. 210.
104. O'Connor, *World Crisis in Oil*, p. 137.
105. P. Collier & D. Horowitz, *The Rockefellers: An American Dynasty* (New American Library, New York, 1976), pp. 207–8.
106. Sampson, *The Seven Sisters*, p. 78.
107. E. Gray, *The Great Canadian Oil Patch* (Maclean-Hunter Limited, Toronto, 1970), pp. 63–64.

108. Ibid., p. 63.
109. J. Richards & L. Pratt, *Prairie Capitalism: Power and Influence in the New West* (McClelland and Stewart, Toronto, 1979), p. 46.
110. Gray, *The Great Canadian Oil Patch*, p. 72.
111. Richards & Pratt, *Prairie Capitalism*, p. 47.
112. Gray, *The Great Canadian Oil Patch*, p. 73.
113. Ibid., p. 76.
114. Ibid.
115. Richards & Pratt, *Prairie Capitalism*, p. 57.
116. Ibid.
117. Ibid., p. 47.
118. Gray, *The Great Canadian Oil Patch*, p. 82.
119. Canada, Director of Investigation and Research, Combines Investigation Act, *The State of Competition in the Canadian Petroleum Industry* (7 vols., Minister of Supply and Services Canada, Ottawa, 1981), vol. 1, p. 186.
120. Collier & Horowitz, *The Rockefellers*, p. 208.
121. Vallenilla, *Oil: The Making of a New Economic Order*, p. 53.
122. Ibid., pp. 55–57.
123. Ibid., p. 57.
124. O'Connor, *World Crisis in Oil*, p. 113.
125. Vallenilla, *Oil: The Making of a New Economic Order*, p. 54.
126. Ibid., pp. 57–58.
127. Jacoby, *Multinational Oil*, p. 41.
128. U.S. Federal Trade Commission, *The International Petroleum Cartel*, p. 130.
129. Ibid., p. 71.
130. Gulf sold its shares in IPC in 1931 to free itself from the agreement.
131. U.S. Federal Trade Commission, *The International Petroleum Cartel*, pp. 130–31.
132. N. Robertson (ed.), *Origins of the Saudi Arabian Oil Empire, Secret U.S. Documents, 1923–1944* (Documentary Publications, Salisbury, North Carolina, 1979), p. i.
133. Ibid., p. 19.
134. Ibid., p. 21.
135. Ibid.
136. Ibid., p. 1.
137. Ibid., p. 25.
138. Ibid., pp. 52–53.
139. Ibid., p. 55.
140. Ibid., p. 48.
141. Ibid., pp. 49–50.
142. Ibid., p. 84.
143. Ibid., pp. 83–84.
144. Ibid., p. 88.
145. Ibid., p. 97.
146. U.S. Federal Trade Commission, *The International Petroleum Cartel*, p. 116.
147. Ibid., pp. 114–16.

CHAPTER FIVE

# The Relationship between the Oil Industry and the American State—Post-World War II Era

World War II, even more dramatically than World War I, illustrated the importance of oil in modern warfare. Both the Allied and Axis forces were highly mechanized, with large fleets of aircraft, trucks, tanks, self-propelled artillery, and ships. All this equipment consumed enormous quantities of oil.

Many key battles were fought over oil. The German drive to Stalingrad was motivated partly by the desire to cut off Red Army oil shipments coming up the Volga from Iran, and partly by an attempt to secure the oil-rich Caucasus for the Third Reich. The Japanese invasion of Indonesia was likewise prompted by the need to obtain that area's oil. The massive Allied air raids on Ploesti in Rumania were aimed at depriving Germany of its major oil source. These are but a few of the many examples of oil's strategic role in the war.

Within the United States, oil policy was a matter of paramount concern. As in World War I, the U.S. supplied the Allies with most of their oil. The U.S. furnished nearly five billion of the seven billion barrels consumed by the Allies between 1941 and 1945.[1] Also, as in World War I, the U.S. feared that it would soon deplete its own oil resources. As an example of this concern, there is the letter from William C. Bullitt, under-secretary of the navy, to President Roosevelt in June 1943, stating:

> *The crude petroleum reserves of the United States, at either present rates of withdrawal or peace-time rates, will be totally exhausted in about 14 years*—unless new reserves of major importance should be discovered. Our leading petroleum geologists express the opinion that there are no known major structures the development of which would make more than a minor change in the situation.[2] (Italics as in original)

Bullitt then raised the question of finding secure future oil supplies.

Following Bullitt's letter, top government leaders engaged in an extensive discussion of postwar oil policy. Before analysing in detail the postwar evolution of this policy, it is first necessary to examine the environment in which this policy was formulated. The war drastically altered the world power structure, and the United States emerged from the war as the dominant power in the world.

## Dominance of the United States

The United States, unlike most of the belligerents, escaped the ravages of war. Together with its northern neighbour, Canada, it emerged from the struggle with an enhanced economic base. Gross National Product, for instance, grew in real terms by 49 per cent between 1939 and 1946. Business investment, the key to increased capacity, expanded much more—by 88 per cent.[3] Furthermore, U.S. casualties were relatively light, both in comparison with the other belligerents, including Canada, and compared to U.S. casualties in previous wars. Though the U.S. suffered over one million casualties, including more than 400,000 deaths, the ratios of both casualties and deaths to armed forces personnel were about equal to the ratios in World War I, approximately 7 per cent for total casualties and 2.5 per cent for deaths,[4] despite the fact that the United States was engaged in the second war for a far longer period (forty-six months for World War II, compared to nineteen months for World War I). Thus the comparative losses of the United States were smaller in the later conflict.[5] Furthermore, since the U.S. was neither invaded nor bombed, it suffered few civilian casualties.

At the end of the war the U.S. thus had an intact military machine, an effective civilian labour force, and an enlarged industrial base. In addition it had acquired a significant lead in technology. It developed and perfected many new weapons, including the atom bomb, on which it had a monopoly. Given its superb industrial and military machine and its possession of the bomb, the U.S. was in a position to impose a *Pax Americana* on most of the world.

Only in Eastern Europe, the Soviet Union, and the Communist-held areas of China was the U.S. unable to assume a dominant position. Despite the immense devastation these lands suffered, they still managed to retain an effective military capacity. The Soviet Union, in particular, emerged from the war as a major military power. Though relatively weak compared to the U.S., it was nevertheless strong enough to keep the U.S. from dominating these areas.

The refusal of the Soviets to accept a *Pax Americana* was one of the factors that led to the Cold War. A detailed analysis of the origins of the Cold War is beyond the scope of this book.[6] Nevertheless, it is useful to question the conventional wisdom that the Cold War represented a defensive American response to the threat of Soviet expansionism. American policy in that period was anything but defensive. It followed the traditional policy of acquisition without annexation, *i.e.*, the policy of Carl Schurz. Under the guise of defending the "Free World," the U.S. was able to exercise a dominant influence over most non-Communist nations.

*Acquisition without Annexation*

The U.S. acquired military bases throughout the world. Schurz, it should be recalled, advocated such a policy as a means of obtaining economic dominance. By 1968 the U.S. had bases and facilities in sixty-four different countries, as can be seen in Table 5.1. It has been noted that this was far more than in any previous period in U.S. history, including World War II. The "U.S. armed forces in the 1920s," Magdoff pointed out, "were stationed in only three countries abroad. During the Second World War, U.S. armed forces were to be found in thirty-nine countries."[7] Another interesting aspect is that a very large number of facilities were located in areas like Latin America, Africa, South Asia, and East Asia—areas to which the Soviet Union could not possibly have posed a serious military threat. But the presence of these facilities did create a favourable climate for U.S. trade and investment. American forces in an area presumably ensured political stability, a

**Table 5.1: Geographic Distribution of U.S. Bases and Facilities, 1968**

| *Region* | *Number of Countries in Which U.S. Armed Forces Were Stationed* |
|---|---|
| Latin America | 19 |
| Europe | 13 |
| Africa | 11 |
| Near East and South Asia | 11 |
| East Asia (including Australia) | 10 |
| Total | 64 |

Source: H. Magdoff, *The Age of Imperialism, The Economics of U.S. Foreign Policy* (Modern Reader Paperbacks, New York and London, 1969), p. 42.

necessary ingredient of any large-scale international investment program.

*The Export of Capital*

One of the cornerstones of U.S. foreign economic policy was the encouragement of private investment abroad. This has been a consistent aim of all post–World War II U.S. governments, irrespective of party. President Truman, for instance, stated in his 1949 inaugural address that "we should foster capital investment in areas needing development."[8] President Eisenhower in 1953 said that "a serious and explicit purpose of our foreign policy [is] the encouragement of a hospitable climate for investment in foreign nations."[9] Dean Rusk, secretary of state under Presidents Kennedy and Johnson, told a congressional committee hearing that foreign countries should

> create conditions which will be attractive to the international investor, the private investor. So our influence is used wherever it can be and persistently, through our Embassies on a day-to-day basis, in our aid discussion and in direct aid negotiation, to underline the importance of private investment.[10]

Since the bulk of private foreign investment then, and now, came from the United States, Dean Rusk was really talking about American private investment. His remarks, like those of Presidents Truman and Eisenhower, were essentially no different from those of President Taft in his famous "dollar diplomacy" speech.

FOREIGN AID

Dean Rusk moreover mentioned an additional policy tool, the tool of foreign aid. In Taft's day, the U.S., which was still in a net debtor position, was not able to dispense large sums of money for the purpose of opening up trade and investment opportunities for its corporations. In contrast, by the end of World War II the U.S. had become the world's leading creditor and was thus in a position to use aid as a key policy tool. As President Kennedy stated in 1962, "foreign aid is a method by which the United States maintains a position of influence and control around the world. . ."[11]

By pouring dollars into a dollar-short world, the U.S. aid program created markets for its exports and outlets for its investment capital. In early postwar years the aid outlays enabled the U.S. to amass huge balance of trade[12] surpluses. As Table 5.2 shows, aid expenditures in the first ten postwar years, the period in which the dollar shortage was most severe, equalled $53 billion, a sum only $3 billion less than the trade

**Table 5.2: Relationship between U.S. Foreign Aid Expenditures and the Balance of Trade Surplus, 1946–55**

| | *Net Aid Expenditures*[a] ($ Billions) | *Balance of Trade Surplus*[b] ($ Billions) | *Aid as Percentage of Surplus* (%) |
|---|---|---|---|
| 1946 | 5.4 | 6.7 | 81 |
| 1947 | 6.2 | 10.1 | 61 |
| 1948 | 5.3 | 6.0 | 88 |
| 1949 | 5.8 | 5.4 | 107 |
| 1950 | 4.1 | 1.5 | 273 |
| 1951 | 4.6 | 4.3 | 107 |
| 1952 | 5.0 | 5.1 | 98 |
| 1953 | 6.3 | 5.5 | 114 |
| 1954 | 5.0 | 5.8 | 86 |
| 1955 | 4.8 | 5.4 | 89 |
| Total, 1946–55 | 52.5 | 55.8 | 94 |

a. Includes civilian and military grants and credits less loan repayments.
b. Merchandise exports less merchandise imports. Includes exports made under military grants programs.

Source: U.S. Department of Commerce, *Balance of Payments Statistical Supplement*, rev. ed. (U.S. Government Printing Office, 1963), pp. 3, 154–57.

surplus. Had this supply of dollars not been forthcoming, the dollar-hungry world would not have been able to purchase the goods produced by American corporations, and the U.S. could not have acquired such huge surpluses.

The transfer of dollars also created a favourable investment climate, as U.S. corporations investing abroad were certain that they would receive dollars for their products sold abroad. In addition this aid was often used to provide the infrastructure necessary to induce private investment. Furthermore, specific aid programs, like Public Laws 480 and 128, allowed American firms to borrow the foreign currencies the U.S. had accumulated in less-developed countries to make investments in those countries. The government also provided investment guarantees generally covering losses up to 75 per cent of the investment. Finally, Congress in 1962 passed the Hickenlooper Amendment to the Foreign Assistance Act, an amendment which provided for mandatory cut-offs of aid to any country expropriating U.S.-owned property unless that country made immediate and full payment to its owners. Perhaps more than any other piece of legislation, the Hickenlooper Amendment bared the link between foreign aid and investment.

*International Monetary Regime*

The international monetary regime also facilitated the flow of capital abroad. Under the rules of the International Monetary Fund (IMF), adopted at the insistence of the United States at the Bretton Woods Conference of 1944, the U.S. dollar became, along with gold, the means of settling international accounts. This special status of the dollar removed normal restraints on U.S. international investments: American firms seeking assets abroad did not have to acquire either gold or foreign currency to purchase them; they purchased them with U.S. dollars. Because of the widespread acceptability of the dollar, they were able to continue expanding abroad even after the United States began incurring large balance of payments deficits in the 1950s. As long as foreign governments and banks were willing to hold dollar reserves, American companies were free to invest wherever they pleased. Foreigners, on the other hand, had to worry about the costs of acquiring either U.S. dollars

**Table 5.3: Per Cent Changes in Values of Key Economic Indicators, Selected Years**

| | *1946–55* | *1955–65* | *1965–75* |
|---|---|---|---|
| Foreign Investment | 168 | 155 | 152 |
| GNP | 91 | 72 | 122 |
| Domestic Investment | 96 | 65 | -47 |
| Corporate Investment[a] | 103 | 112 | 38 |
| Exports | 22 | 84 | 304 |

a. Net investment of the corporate sector excludes residential and agricultural investments. The latter are included in domestic investment, which encompasses all investment in the economy.

Sources: *Historical Statistics of the United States, Colonial Times to 1957: A Statistical Abstract Supplement* (U.S. Government Printing Office, Washington, 1960), p. 565; M. Wilkins, *The Maturing of Multinational Enterprise: American Business Abroad from 1914 to 1970* (Harvard University Press, Cambridge, Mass., 1974), p. 329; *Economic Report of the President Transmitted to the Congress January 1980* (U.S. Government Printing Office, Washington, 1980), pp. 203, 216, 219, 222, 316, and 321; and *Economic Report of the President Transmitted to the Congress January 1979* (U.S. Government Printing Office, Washington, 1979), p. 298.

or gold before they could invest abroad. Thus, as a result of both the U.S. aid programs and the status of the dollar, U.S. direct foreign investment grew at a rapid rate.

As can be seen in Table 5.3, in each of the three postwar decades the book value of foreign investment rose at a faster rate than several other key indicators, with only one exception. That exception occurred in the third decade when merchandise exports expanded at a faster pace than foreign investment. During this decade the U.S. dollar lost much of its status as an international currency. Because of the persistent balance of payments deficit, the dollar shortage of the 1940s and the early 1950s had turned into a dollar glut. By the late 1960s many governments and banks were refusing to add to their dollar reserves. Instead of dollars they demanded gold or other hard currencies. This loss of acceptability forced a devaluation of the dollar, with the result that foreign investments became more costly to American firms. At the same time American goods became cheaper to foreigners. Thus merchandise exports grew at a faster rate than foreign investments. Nevertheless, the growth in foreign investments was significantly higher than in other key indicators like GNP, Net Private Domestic Investment, and Net Corporate Investment.

There is little doubt that the export of capital played a major role in the American economy in the postwar years. Not only did foreign investment expand much faster than domestic investment but also the earnings from foreign investment became a key element in the United States' balance of payments position. As Table 5.4 indicates, income from foreign investments became an increasingly larger portion of the balance on goods and services[13] and has actually exceeded the balance since the mid-1960s. The increase in investment income partially offset the decline in the trade balance, which during the 1970s turned from a surplus to a deficit. Thus, by the 1970s the American Empire reached the same position as the British Empire at the turn of the century, when Britain had become a net importer of merchandise and relied on income from foreign investments to balance its international accounts. The United States has similarly become dependent on the income from its overseas investments.

### *U.S. Trade Balance*

As Table 5.5. shows, the United States' balance of trade surplus reached a peak in the early postwar years and has followed a secular downtrend ever since. By the 1970s the balance of trade surplus had turned into a

**Table 5.4: Relationship between Net Income on Foreign Direct Investment and the Balance on Goods and Services, 1946–78**

| *Years* | *Income*[a] (Annual Average) ($ Billions) | *Balance*[b] (Annual Average) ($ Billions) | *Income as Percentage of Balance* (%) |
|---|---|---|---|
| 1946–49 | 1.3 | 8.2 | 16 |
| 1950–54 | 2.3 | 2.7 | 85 |
| 1955–59 | 3.5 | 3.9 | 90 |
| 1960–64 | 5.0 | 6.8 | 74 |
| 1965–69 | 7.1 | 5.4 | 132 |
| 1970–74 | 12.8 | 5.3 | 242 |
| 1975–78 | 27.7 | 3.7 | 749 |

a. Investment income payments to Americans by foreigners minus similar payments by Americans to foreigners. Included in investment income are interest, dividends, royalties, and service fees. The latter two are classified by the U.S. Department of Commerce as "other service income." In this table, they are included in investment income because they are related to the technology transfer process which normally accompanies international investment.
b. The balance on goods and services encompasses the trade balance plus the balance on services like insurance, travel, freight, and investment income.

Source: *Economic Report of the President Transmitted to the Congress January 1980* (U.S. Government Printing Office, Washington, 1980), p. 316.

**Table 5.5: United States Balance of Trade, 1946–78**

| *Years*\ | *Balance* (Annual Average) ($ Billions) |
|---|---|
| 1946–49 | 7.0 |
| 1950–54 | 2.9 |
| 1955–59 | 3.7 |
| 1960–64 | 5.4 |
| 1965–69 | 2.8 |
| 1970–74 | −2.1 |
| 1975–79 | −16.3 |

Source: *Economic Report of the President Transmitted to the Congress January 1980* (U.S. Government Printing Office, Washington, 1980), p. 316.

deficit. The U.S. deficit differed, however, from the British one in one significant respect: while the British deficit was caused by the imports of agricultural goods and raw materials, the American one was caused by the importation of manufactured goods. A portion of these manufactured goods was produced by the overseas facilities of American corporations. Thus, in the development of the American Empire the export of capital first tended to replace the export of goods and later to displace the domestic production of goods.

It is significant that the U.S. balance of trade deficit was not caused by the rise in oil prices. The first deficit occurred in 1971, a full three years before the fourfold oil price increase of 1974. In 1971 manufactured imports accounted for approximately two-thirds of the import bill.[14] The oil bill was about 8 per cent.[15] By 1978, when oil accounted for almost one-quarter of all imports, manufactured imports still accounted for more than one-half.[16]

The rise in imports from U.S.-owned facilities abroad was directly attributable to the growth in investment in manufacturing. As can be seen in Table 5.6, more than 40 per cent of all direct investment in 1970 was in manufacturing. Manufacturing, in comparison, accounted for only one-third of investment at the war's end.[17] This predominance of manufacturing investment stands in contrast to earlier years, when manufacturing ranked well below mining and smelting (see Table 2.4).

**Table 5.6: Distribution of U.S. Direct Private Investments by Industry, 1970**

| *Industry* | *Investment* ($ Billions) | (%) | *Per Cent Change* (1946–70) |
|---|---|---|---|
| Manufacturing | 32.3 | 41 | 1,246 |
| Petroleum[a] | 21.7 | 28 | 1,450 |
| Trade | 6.6 | 8 | 1,000 |
| Mining and Smelting | 6.2 | 8 | 675 |
| Public Utilities[b] | 2.9 | 4 | 123 |
| Other[c] | 8.5 | 11 | 1,114 |
| Total | 78.2 | 100 | |

a. Includes production, refining, marketing, transportation, and shipping.
b. Includes transportation.
c. Includes agriculture, sales organizations, banks, and insurance companies.

Source: M. Wilkins, *The Maturing of Multinational Enterprise: American Business Abroad from 1914 to 1970* (Harvard University Press, Cambridge, Mass., 1974), p. 329.

**Table 5.7: Geographical Distribution of U.S. Direct Private Foreign Investment, 1970**

| *Region* | *Total* ($ Billions) | (%) | *Manufacturing* ($ Billions) | (%) | *Petroleum*[a] ($ Billions) | (%) |
|---|---|---|---|---|---|---|
| Europe | 24.5 | 31 | 13.7 | 42 | 5.5 | 25 |
| Canada | 22.8 | 29 | 10.1 | 31 | 4.8 | 22 |
| Latin America | 14.8 | 19 | 4.6 | 14 | 3.9 | 18 |
| Asia | 5.6 | 7 | 1.5 | 5 | 3.0 | 14 |
| Africa | 3.5 | 4 | 0.5 | 2 | 2.1 | 10 |
| Oceania | 3.5 | 4 | 1.8 | 6 | 0.7 | 3 |
| Unallocated | 3.5 | 4 | 0.1 | b | 1.7 | 8 |
| Total | 78.2 | 100[c] | 32.3 | 100 | 21.7 | 100 |

a. Includes production, refining, marketing, transportation, and shipping.
b. Less than one-half of one per cent.
c. Because of rounding, the total may not equal the sum of the column.

Source: M. Wilkins, *The Maturing of Multinational Enterprise: American Business Abroad from 1914 to 1970* (Harvard University Press, Cambridge, Mass., 1974), p. 330.

As in the past, most manufacturing investment was concentrated in the developed countries. The data in Table 5.7 show that almost three-quarters of manufacturing investment was located in Europe and Canada. Only three-fifths of total investment was placed in these areas. This pattern was similar to the one prevailing in 1914 (see Tables 2.5 and 2.6). The typically colonial pattern had not changed.

### CANADA

What had changed was the increasing influence the U.S. corporations exercised over the developed countries. In many countries, and especially in Canada, U.S. corporations controlled key sectors of the economy. During the post–World War II years, American investment in Canada continued the trend, begun after World War I, of displacing British capital as the main source of foreign funds. As can be seen in Table 5.8, U.S. investments accounted for only 22 per cent of all foreign investments in Canada in 1913. By 1946 this share had risen to 72 per cent and in 1965 it had reached almost 80 per cent.

What is more important from the point of view of Canadian development is that this geographical shift was accompanied by a qualitative shift. American *direct* investment was substituted for British *portfolio* investment. As Table 5.8 shows, direct investment made up only one-fifth of all foreign investment in Canada in 1913. By 1946 this share had risen to almost two-fifths, and by 1965 it was approaching three-fifths.

The significance of this shift lies not only in the movement of economic decision-making centres from Canada to the U.S. but also in its implications for perpetual peonage. Portfolio investment, which has consisted mainly of debt investment, has a termination date. Like a mortgage, it is paid off when due, leaving the borrower free to use his property in any way he sees fit. Debt instruments thus provide the means through which the borrower can free himself from the debt burden.

Direct investment, on the other hand, provides no such mechanism. As long as the foreign investor refuses to sell his asset, the market mechanism will permit him to earn income from his investments from here to eternity. Under these circumstances the only way a recipient country can liquidate its foreign obligations is to go outside the market place and to have the government use its sovereign power to force liquidation. As Aitken has noted:

> Direct investments typically involve the extension into Canada of organizations based in other countries; these organizations establish themselves in

**Table 5.8: Foreign Capital Invested in Canada, 1913, 1946, 1965**

| | *1913* ($ Millions[a]) | (%) | *1946* ($ Millions[a]) | (%) | *1965* ($ Millions[a]) | (%) |
|---|---|---|---|---|---|---|
| Total | 3,850 | 100 | 7,178 | 100 | 29,507 | 100 |
| U.S. | 835 | 22 | 5,157 | 72 | 23,305 | 79 |
| U.K. | 2,818 | 73 | 1,668 | 23 | 3,498 | 12 |
| Other | 197 | 5 | 353 | 5 | 2,704 | 9 |
| Direct | 770 | 20 | 2,826 | 39 | 17,208 | 58 |
| Portfolio | 3,080 | 80 | 4,352 | 61 | 12,299 | 42 |
| Direct | 770 | 100 | 2,826 | 100 | 17,208 | 100 |
| U.S. | 520 | 68 | 2,428 | 86 | 13,940 | 81 |
| U.K. | 200 | 26 | 335 | 12 | 2,013 | 12 |
| Other | 50 | 6 | 63 | 2 | 1,255 | 7 |
| Portfolio | 3,080 | 100 | 4,352 | 100 | 12,299 | 100 |
| U.S. | 315 | 10 | 2,729 | 63 | 9,365 | 76 |
| U.K. | 2,618 | 85 | 1,333 | 30 | 1,485 | 12 |
| Other | 147 | 5 | 290 | 7 | 1,449 | 12 |

a. Canadian dollars.

Calculated from: K. Levitt, *Silent Surrender: The Multinational Corporation in Canada* (Macmillan of Canada, Toronto, 1970), p. 67.

**Table 5.9: Degree of Non-Resident Majority Ownership of Corporations in Canada, Selected Industries, 1967**

| *Industry* | *Assets of Foreign-Controlled Corporations as a Percentage of Industry Assets* |
|---|---|
| Mining | 60.0 |
| Mineral Fuels | (81.7) |
| Other Mining | (50.0) |
| Manufacturing | 56.7 |
| Tobacco Products | (83.6) |
| Rubber Products | (92.4) |
| Textile Mills | (49.6) |
| Primary Metals | (55.6) |
| Machinery | (71.9) |
| Transport Equipment | (86.2) |
| Electrical Products | (65.7) |
| Non-Metallic Mineral Products | (47.1) |
| Petroleum and Coal Products | (99.6) |
| Chemicals and Chemical Products | (83.0) |

Source: Canada, *Report of the Royal Commission on Corporate Concentration* (Minister of Supply and Services Canada, Ottawa, 1978), p. 191.

> Canada for purposes of their own and bring with them their own business practices, their own methods of production, their own skilled personnel, and very often their own market outlets. If all Canadian borrowings from other countries were to cease tomorrow, these direct investment organizations would continue to exist and function.[18]

This permanent, direct investment was, furthermore, concentrated in a few key industries, industries which can be viewed as essential to the development and growth of the country. As is shown in Table 5.9, in 1967 foreign corporations, mainly American, controlled 60 per cent of the assets in mining and 57 per cent of the assets in manufacturing. Such key manufacturing industries as rubber products, machinery, transport equipment, electric products, petroleum and coal products, and chemicals and chemical products were dominated by U.S. interests. This control by U.S. corporations gave the U.S. government the potential to control the entire Canadian economy.

This control in Canada, and to a lesser extent in other developed countries, raised American power and influence to unprecedented heights. By the end of the 1960s the American Empire had expanded beyond the wildest dreams of Carl Schurz.

## Oil in America's Expansion

The oil industry played a key role in this expansion. Though investment in manufacturing in 1970 was greater than in oil, the rate of growth in oil investments during the postwar period was, as Table 5.6 shows, greater than for any other industry. As has already been indicated in Tables 3.5 and 4.7, oil's growth rate in earlier periods was below that of other industries. The rapid expansion of oil investments in the postwar era was directly related to the need to consolidate and secure the growing American Empire.

*Bullitt's Proposals*

The roots of this rapid expansion lay in World War II when, as already mentioned (see page 77), William Bullitt drafted his 1943 warning to President Roosevelt. He, like most other U.S. leaders, was aware of the new role America was to play in the postwar world. Perturbed by the prospect that an oil-deficient United States might not be able to play this role, Bullitt told the president:

> We can and must act to provide additional reserves outside the continental limits of the United States. Oil is and will remain for some decades a vital element both for our national defense and our national economy. *To acquire petroleum reserves outside our boundaries has become, therefore, a vital interest of the United States.*[19] (Italics as in original)

What Bullitt proposed was the establishment of a government-owned Petroleum Reserves Corporation which would "acquire for the Government of the United States a controlling interest in proved foreign oil reserves."[20] As a first step he suggested that the Corporation "give immediate attention to the acquisition of a controlling interest in the concessions now held in Saudi Arabia by the California Arabian Standard Oil Company,"[21] which was then owned jointly by Standard Oil of California and the Texas Company.

Interestingly enough, Bullitt drew a parallel between his proposed Corporation and the British government's acquisition of stock in Anglo-Iranian. He argued that if the U.S. government did acquire the controlling interest in the Saudi Arabian company and rendered to it "support of the same effective nature that the British Government has given and is giving the Anglo-Iranian Oil Company, the controlling interest in which is owned by the British Government—it might be possible to retain the concession in American hands and develop it successfully."[22]

He emphasized why he favoured this British approach rather than the

American one of giving diplomatic backing to private U.S. companies operating abroad. In rebutting a suggestion that the government, instead of purchasing stock in the Saudi company, buy an interest in reserves of about 300 million barrels, he stated:

> To accept any such proposal would be to relapse into the "dollar diplomacy" of a dead era. The Government of the United States would be undertaking to back up with all its diplomatic and other resources private capitalists whose concessions will be relatively valueless without government support. In exchange for the reservation of 300,000,000 barrels of oil, our Government would be virtually guaranteeing private owners in the possession of 20 billion barrels of oil. *No such policy should be countenanced by our Government.*[23] (Italics added)

Finally Bullitt urged that the Corporation should not limit itself to Saudi Arabia. He wanted it to "explore all other possibilities in all quarters of the earth."[24] Given only a fourteen-year supply in the United States, he argued that with the acquisition of foreign oil reserves the U.S. "domestic reserve should be kept in the ground as a strategic reserve and that our current needs should be filled from sources of oil outside the continental limits of the United States."[25] Thus, while ostensibly denouncing "dollar diplomacy," he advocated a policy which would denude the world of its oil in order to serve the interests of the United States. The world's oil henceforth was to belong directly to Uncle Sam. Oil imperialism was to replace dollar diplomacy.

While the U.S. government gave serious consideration to Bullitt's import proposal, it was never able to implement it. The reason lay in the private ownership of domestic oil resources. Those companies with large domestic reserves were unwilling to forego income for long periods of time in order to permit imports to serve the U.S. market. The loss of present income for some uncertain future income would have reduced the present values of their reserves. No private company would be willing to incur such a loss unless it received compensation in return. The government, quite understandably, was unwilling to pay the private producers the large sums necessary to induce them to shut in their wells. The only other alternative would have been nationalization. That was both costly and politically unacceptable.

In retrospect, Bullitt's proposal was a sensible one for the United States. Had United States wells been shut in, Middle Eastern and other foreign oil would have been depleted at a much faster rate. Furthermore, the United States and its allies would not have been vulnerable to any embargo. The untapped reserves of the United States would have been able to offset any severe shortages. OPEC would probably never have been formed. If it had, its bargaining power would have been seriously weakened as long as the United States was willing to throw its reserves on the market to prevent prices from rising. But in a capitalist state, where private interests must prevail, it is often difficult, if not impossible, to act in the national interest, especially if the national interest seriously conflicts with the interests of powerful private parties.

### *Petroleum Reserves Corporation*

While Bullitt's import proposal was not adopted, his suggestion for a Petroleum Reserves Corporation was. The Petroleum Reserves Corporation was created by executive order, *i.e.*, by the executive branch and not by Congress, in July 1943. Its board of directors consisted of the secretary of state, the secretary of war, the secretary of the navy, the secretary of the interior, and the director of the Office of Economic Warfare.[26] The Corporation was obviously a top-level body run by some of the most powerful men in the government.

Despite protracted negotiations the Corporation failed to acquire control of the Saudi company. Initially the two partners agreed to let the Corporation buy one-third of the shares. Later the Texas Company changed its mind, claiming that the Corporation did not offer it enough for the shares.[27] Eventually the scheme was abandoned and the Corporation was dissolved in 1944.[28]

It is interesting to note that this high-level corporation was unable to force either oil company to sell its shares to it in order to protect the presumed national interest. This incident illustrates what the true power relationships in the United States were during the war. It also shows how well entrenched the oil industry had become in the American establishment since the dissolution of the Standard Oil Company in 1911.

### *The Industry's Alternative*

The entire oil industry opposed both the establishment of the Corporation and the acquisition by that corporation of stock in private companies. In 1943 the Petroleum Industry War Council, an organization representing virtually every oil company in the United States, adopted a resolution stating that "oil development can best be handled by private enterprise."[29] It went on to assert that "oil in the hands of nationals of the United States is equally available for national security with oil owned or financially shared in by the Government of the United States."[30] It then concluded that the "Government of the United States should encourage private American enterprise to engage in the development of oil resources abroad."[31]

From these general statements it made some specific points. It, for instance, gave the following estimate of how much foreign reserves the U.S. companies should acquire:

> The combined domestic and foreign oil reserves held by nationals of the United States constitute a smaller proportion of the petroleum reserves of the world than the ratio of United States consumption of petroleum to total world consumption. Action is needed to enlarge the reserves under the stewardship of nationals of the United States . . .[32]

This doctrine, if carried to its logical conclusion, would have given the United States control of most of the world's mineral resources since the United States was then, and still is, the largest single consumer of most raw materials in the world. Presumably mere consumption, no matter how wasteful, justifies the concomitant reserves. Such were the benefits that could be expected from a *Pax Americana*.

The resolution also stressed the need for acquiring many types of facilities throughout the world.

> To enhance national security many and widely distributed sources of products for military supply should be in the hands of United States nationals. Such sources have been developed by private industry and in many areas can only be so developed. Only through distributing facilities as well as sources of supply can these requirements be assured.[33]

This paragraph, it should be noted, speaks of investment in general, not merely oil investments. It is interesting in that it reveals the attitude of an important segment of American business towards the link between economic and military power: it is a relationship of interdependence. Military power creates the conditions necessary for economic expansion. Economic power provides the basis for military power.

This link between the economic and military variables is implied in the following statement:

> The government of the United States can serve the interests of a greater number of its own nationals and at the same time promote the interests of foreign countries by means of a foreign oil policy designed to reduce the political risks inherent in foreign oil operations.[34]

While neither "political risks" nor the designs to reduce them are explicitly defined, one can presume that the most worrisome risk is nationalization and that military intervention can reduce this risk. It is significant that nowhere in the resolution is military intervention explicitly ruled out as a policy instrument for reducing political risks.

*Collado's Proposals*

Once the Petroleum Reserves Corporation had been scuttled, the United States government had no choice but to adopt the main policy lines of the Petroleum Industry War Council. Though State Department officials expressed some reservations, they were forced to accept this policy. In a memorandum dated March 29, 1945 to William Clayton, head of the State Department's Economic Affairs Division and assistant secretary of state, Emilio Collado of the department's Office of Fuel Development questioned whether national security required that Saudi Arabian oil be controlled by American companies. Collado wrote:

> It is not of vital importance to the military security of the United States that Arabian oil remain in the hands of an American company. The Navy has admitted that Arabian oil would not be subject to American strategic control in the next war. The British would be our Allies, and if it could be controlled by them, it would be as good as ours.[35]

There are two points of interest in this statement. One is Collado's acceptance of the inevitability of another war even before World War II had come to its end. The other is the willingness to let the British have custody of Arabian oil. The latter point represented a break with the traditional State Department's view of Britain. Recognizing this, Collado commented:

> Most foreign service officers who have served in the Middle East have an inferiority complex vis-à-vis the British, resulting from years of being outsmarted and operating without much interest or backing from home. Arabian oil is the greatest opportunity they have ever had to stimulate interest in the Middle East and U.S. government backing against their long-time adversaries, the British. They will not acknowledge that cat and dog fights with the British in the Middle East back yard have now become foolish and futile, since the British can be handled quite simply by a firm hand in Washington.[36]

Collado, who was later to leave the State Department to become an executive in Standard Oil (New Jersey), thus explicitly recognized the change in international power relations brought about by the war. Britain had been reduced to a dependency of the United States.

Collado further argued that there were distinct advantages in giving Saudi oil to a British firm. Among them would be the ability to implement a modified version of Bullitt's import plan by having Middle East oil replace Western Hemisphere oil in the European market.

> Arabian oil, if properly handled, could contribute to the military security of the western hemisphere. If developed and marketed in Europe, replacing western oil that might be marketed there, western reserves might be conserved to the extent of such replacement. The unpopularity of this procedure with the great portion of the American petroleum industry not represented in the Middle East will make it very difficult for the U.S. Government to pursue an aggressive course in this regard. It is interesting to contemplate that the desired result would more likely be obtained if Arabian oil were in British hands, so that the replacement process in Europe would take place through British competition without any encouragement from the U.S. Government. The U.S. Government could also avoid embarrassment if the American oil company itself [ARAMCO] carried out the replacement without undue assistance from the Government (though the progress of an American company will be slow relative to that of a British company because of the dollar payment difficulty).[37]

**Table 5.10: Estimated Distribution of Gross and Fixed Assets of the Petroleum Industry in the Middle East, by National Ownership and Control, Beginning of 1947 and 1959 (Per Cent)**

| *Country* | *1947* | *1959* |
|---|---|---|
| United States | 40 | 50 |
| United Kingdom[a] | 44 | 18 |
| France | 8 | 5 |
| Netherlands[a] | 8 | 3 |
| Others[b] | — | 24 |
| Total | 100 | 100 |

a. Shell's investments were allocated as follows: 60 per cent to the Netherlands and 40 per cent to the United Kingdom.
b. This mainly represents the holdings of the National Iranian Oil Company. As part of the settlement following the restoration of the Shah, the Iranian government in 1954 took over the properties held by the Anglo-Iranian Oil Company. It also signed an agreement with the consortium known as Iranian Oil Participants Ltd., giving that group production and exploration rights in southern Iran and the right to operate the refinery at Abadan. The national distribution of ownership of the consortium was: United Kingdom, 46 per cent; United States, 40 per cent; Netherlands, 8 per cent; France, 6 per cent. This distribution is included in the nationality figures.

Source: C. Issawi & M. Yeganeh, *The Economics of Middle Eastern Oil* (Praeger, New York, 1962), pp. 59, 175–76.

Despite the advantages, Collado recognized that there would be political difficulties in implementing his proposal. In an earlier memorandum to Clayton, he noted that "it is probable that the State Department would be subjected to severe public criticism if the concession were lost either to Britain or Russia."[38] Thus, despite Collado's misgivings, Saudi oil remained firmly in the hands of the private U.S. companies. Given the government's view of the essential nature of oil, it had no choice but to back these companies in their relations with foreign governments.

### *Expansion in the Middle East*

After rejecting Britain's surrogate role, American policy quite logically concentrated on replacing Britain as the Middle East's dominant oil power. As Table 5.10 shows, the U.S. at the end of World War II was the second oil power in the Middle East, controlling only 40 per cent of the fixed assets there. Britain, despite its weakened condition, retained 44 per cent of the assets. By 1959, the Americans owned 50 per cent, the British, 18. A number of factors contributed to this displacement of Britain.

#### SAUDI ARABIA

The first was the strengthening of the American position in Saudi Arabia. During the war King Ibn Saud became conscious of his country's vast oil potential. Eager to increase his revenues, the king pressed ARAMCO to increase production. In response to his request the owners of ARAMCO, Standard Oil of California and the Texas Company, laid plans to build a pipeline across the country to the Eastern Mediterranean. This pipeline required new markets in Europe, an area in which these companies had only limited marketing facilities. Both the companies and the U.S. government feared that ARAMCO might lose its concession if it could not market this oil.

At this point the State Department intervened. It wanted to hold on to what Walter W. Birge, Jr., American vice consul in Dhahran, Saudi Arabia, described as "probably the largest oil field in the world—America's new frontier."[39] In 1945 the State Department suggested to ARAMCO that it make an arrangement with the American participants in IPC to dispose of its oil. The department proposed that it either sign a marketing agreement with these companies, Standard Oil (New Jersey) and Socony-Vacuum, or have them become partners in ARAMCO.[40] After protracted negotiations, Standard and Socony bought into ARAMCO in early 1947.[41] With the entry of these two partners, the ownership of ARAMCO was distributed as follows: Standard Oil of California, 30 per cent; the Texas Company, 30 per cent; Standard Oil (New Jersey), 30 per cent; and Socony-Vacuum, 10 per cent.[42]

The agreement was advantageous to all sides. The original partners could now go ahead with plans for expansion, knowing that the new partners would find markets for the increased crude production. In addition, Jersey and Socony paid California and Texas $125 million, an amount nearly equal to their investment in Saudi Arabia.[43] Thus the original partners were able to recoup their investment and still retain 60 per cent of the concession. Both Jersey and Socony, for their part, needed increased crude supplies for their growing markets. Jersey, in addition, wanted to make up for the loss of its Rumanian supplies following the nationalization of its properties in that country. Before the war Jersey had supplied much of its European market from Rumania.

That the Saudi venture, involving a joint production and marketing arrangement between four of the U.S.'s largest oil companies, may have violated the antitrust laws deterred neither the companies nor the State Department. These laws apparently were not to stand in the way of the national interest. Nevertheless, the companies did find it convenient to invoke the antitrust laws when these laws corresponded to their interests. The entry of Jersey and Socony into Saudi Arabia broke the "Red Line" agreement, which prohibited any of the IPC partners from obtaining concessions in the territories of the former Ottoman Empire without prior permission of the other partners. In actuality, the IPC as a whole was supposed to obtain any new concession. Since at least two of the partners, CFP and the Gulbenkian interests, objected to the move by Jersey and Socony, the latter two companies notified the other partners that they considered that the "Red Line" agreement was no longer binding upon them.[44] The agreement, they claimed, was "in restraint of trade and contrary to public policy and void and unenforcible in law."[45] The U.S. Federal Trade Commission wryly commented:

> Thus, Standard and Socony followed the red-line agreement for almost 20 years without making any serious objections, but when they wished to obtain an interest in Aramco, they described the agreement as one "in restraint of trade and contrary to public policy and void and unenforcible in law."[46]

By making this admission, Standard and Socony, who had voluntarily signed the "Red Line" agreement, opened themselves up to prosecution under the Sherman Act. However, the U.S. government took no action against them at that time for having signed it. Still unanswered is the question as to what, if any, promises of immunity were obtained from the U.S. government prior to using this antitrust argument as an excuse for abrogating the agreement. In any event the two companies were able to break the contract without suffering any damage. CFP filed a suit against them. They filed a countersuit against CFP. Eventually the whole matter was settled out of court and the "Red Line" agreement passed into history.[47]

There were several reasons why the United States allowed the agreement to be broken with impunity. One was the insistence of Ibn Saud that it be abrogated before Jersey and Socony buy into ARAMCO. The king's objections were outlined in a State Department memorandum, which noted:

> The King, however, is now much disturbed by the possibility that the Red Line Agreement may still be in force, or if not actually in force will still be honored in some way. What he fears is that this will prove an "open door" through which British, French and other oil interests can get a foothold in the development of Arabian oil. The Minister [Asad Al-Faquih, Saudi Arabian Minister to the United States] repeated twice that the King and the Crown Prince were agreed that the development of Arabian oil resources must remain exclusively in American hands. He ended by saying that if the proposed new partners in Aramco could not disassociate themselves from previous commitments, they would not be permitted to buy into the Arabian concession.[48]

Despite the conflict between the king's demands and the traditional U.S. insistence on the "open door," the U.S. was not at all averse to complying with Saud's demands. The reason was that the breaching of the agreement enabled the U.S. companies to retain 100 per cent control of Saudi oil. On the other hand, if the "Red Line" agreement had been observed, then the IPC would have joined the original partners in the Saudi concession. The U.S. companies, it should be recalled, had only a 23.75 per cent interest in IPC. Thus the U.S. ownership of Saudi crude would have been seriously diluted.

### FOREIGN AID

A second factor contributing to the displacement of Britain as the dominant oil power in the Middle East was the U.S. foreign aid program. As discussed earlier in this chapter (see page 000), the U.S. aid program enabled many American firms to enter overseas markets. This was especially true of the Marshall Plan, whose chief beneficiary was the American oil industry and particularly those oil companies with holdings in the Middle East.

Between April 1948 (when the Marshall Plan began operations) and November 1950, the U.S. authorized the countries receiving aid to purchase $384 million of "dollar" oil, *i.e.*, oil produced by U.S. companies in the Middle East.[49] It was this aid that opened up European markets to Saudi and other American-controlled Middle East oil. The importance of this assistance to both the U.S. oil industry and, therefore, the U.S. government cannot be overstressed.

Walter J. Levy, who resigned from Socony-Vacuum to head the oil division of the Marshall Plan, reported that without the plan "the American oil industry in Europe would already have been shot to pieces" and that the "ECA [Economic Cooperation Administration, the official name of the Marshall Plan] has maintained outlets for American oil in Europe... which otherwise would have been lost." He further stated that the "ECA does not believe that Europe should save dollars by driving American oil from the European market."[50] Levy, it should be noted, became the U.S. oil industry's leading consultant after he left the ECA.[51]

In addition to paying for imported American-owned oil, the ECA helped the U.S. oil industry by controlling refinery construction. Before World War II, Europe had little refining capacity. Its imports consisted mainly of products. With the development of the postwar dollar shortage the European governments began a refinery construction program to save foreign exchange. The savings from importing crude, rather than refined products, more than compensated for the foreign exchange costs of refinery construction.[52] The ECA, however, feared that these refineries might be used to import non-dollar crude and tried to limit their construction by restricting aid disbursements. Testifying before a congressional committee in 1949, Dr. Oscar Bransky, acting chief of the ECA's petroleum branch, described ECA oil policy as follows:

> Under the stimulus of the shortage, the oil companies, British and American, both operating world-wide, had made plans in the postwar period to expand rapidly their crude-oil production and refinery capacity. The companies succeeded so well in carrying out their plans that by late 1948 supplies of many products were approximately in balance.... There was no question that a large amount of expansion was justified in order to meet the expanding world demand, and there was also no question that it was ECA's assignment under a Congressional mandate to help the European countries save dollars where possible. Oil is the largest single item in the dollar budget

> of the participating countries, and savings of dollars on oil imports thus ranked high on the list of priorities of these countries.
>
> But at the same time broad considerations of national policy made it essential that ECA do its best to prevent a degree of overexpansion that, because of the dollar shortage, would seriously reduce the market for American-owned oil and thus jeopardize American-owned concessions in foreign lands.[53]

Dr. Bransky, who came to the ECA from Standard Oil (Indiana),[54] presumably worried that the desire of European countries to reduce their imports of dollar oil would lower the demand for that oil and thereby endanger the U.S. concession in Saudi Arabia since King Ibn Saud was then demanding that ARAMCO increase its production. Western Europe was the only logical market for this incremental production. If this market were pre-empted by the British, Dutch, and French companies, ARAMCO could lose its concession. To prevent this from happening the ECA was willing to increase the dollar cost of its program even though it had a congressional mandate "to save dollars where possible." In the words of Levy, where "vital United States interests are at stake...the mere dollar-saving approach is too narrow in the case of oil."[55] Levy, according to Bransky, refused to finance the refinery program developed by the European countries.[56] Furthermore, he advised these countries that the "ECA would be reluctant to sponsor export licences for items of petroleum equipment."[57] Such equipment was then in short supply. He finally induced the U.S. government to place a qualification on authorizations of machinery procured under the aid program so that "such authorizations could not be used for petroleum equipment."[58]

As a direct result of ECA's procurement and refinery policies, the Middle East displaced the Western Hemisphere as the chief source of Europe's oil. In 1947, before the start of the ECA, Europe received 43 per cent of its crude from the Middle East. In 1948, the first year of the Marshall Plan, the Middle East's share rose to 66 per cent. By 1950, it had grown to 85 per cent.[59] Europe had, to a significant degree, become dependent upon American-produced Middle Eastern oil. The full implications of this dependence did not become apparent until some twenty years later.

Before discussing these implications, we should examine some of the more immediate effects of this policy. It substantially enriched the oil companies with Middle East reserves. Their profits were so large that they even enraged the U.S. ambassador to Egypt. In a "restricted" message dated November 16, 1948, to the State Department, Ambassador Griffis took exception to a statement by the president of ARAMCO that the oil companies in the Middle East were financing a junior Marshall Plan there without any cost to the U.S. taxpayer. His message stated:

> On the contrary, information at this end points to the vital conclusion that (1) since the U.S. taxpayer supports the European Recovery program, (2) since ERP [European Recovery Program, another name for the Marshall Plan] oil purchases are 48% derived from U.S. oil companies' Middle East sources (may be 90% by 1951), and (3) since companies sell this oil at prices based on production costs in U.S.A. which are 100 to 500 percent above Middle East production costs, it follows that via the exceptional profits derived from these excess prices, the U.S. taxpayer is contributing millions towards financing Middle East investments mentioned above.[60]

The point about the huge profits on Middle East oil was well founded. Under the basing point pricing system developed by the international oil companies, the price of Middle East oil was based on high-cost Texas oil. The oil companies charged this high price to the ECA while at the same time they charged their affiliates in the United States a lower price. In 1948 Middle East oil began to flow to the United States in substantial quantities. At that time, the "net back price" of Middle East oil, *i.e.*, the New York market price less the U.S. tariff less the tanker charges from the Middle East to New York, was $1.28 a barrel. The oil companies, at the same time, were charging the ECA $2.03 a barrel for Middle East oil shipped to Europe.[61] Throughout the entire period of the Marshall Plan there were similar discrepancies between the price that the oil companies charged the ECA and the price they charged other customers.

In 1952 the ECA filed suit against the two original partners in ARAMCO, Standard Oil of California and the Texas Company, claiming overcharges of $100 million.[62] This suit, like so many similar actions against large, multinational corporations, dragged on in the courts for years. In 1956 the U.S. government amended its complaint and demanded $225 million in damages.[63] In 1957 Judge Thomas Murphy dismissed the government's complaint in a decision which, according to Engler, was based on "a legalistic analysis that skirted most of the economic realities governing the industry."[64]

Notwithstanding Engler's comments, it is questionable whether the government had a case, given the ambiguity of the ECA procurement regulations, many of which were drawn up when Levy was the oil administrator. Frank is undoubtedly quite correct in his observation that

> the Government's case was extremely weak. . . . The great bulk of the shipments made by Socal [Standard Oil of California] and Texas Company to North America were clearly intra-company transactions and exchanges, which could hardly be considered open market sales. It may have been bad public relations to bill these at a figure below the price prevailing for Eastern Hemisphere sales, and perhaps also an offense against the spirit of ECA policy. But it hardly constituted a violation of the statute or the regulations, which were the only provisions legally binding on the participating suppliers.[65]

Questions can also be raised as to the zeal with which the government prosecuted its case. The original suit was filed when Truman was president. It continued under the Eisenhower regime, which had very close links to the oil companies. The actions of the government's attorneys left much to be desired. As Frank has commented, the "ECA had little if any basis for a claim, and would most likely have lost the case, even if it had been handled brilliantly by the Government's attorneys, which was clearly not the case."[66] One certainly has a right to wonder whether the whole prosecution was merely a charade which the government, for political reasons, felt compelled to act out but had no real intention of pursuing.

Wherever the truth may lie, Ambassador Griffis's comment that the U.S. taxpayer, through the Marshall Plan, was "contributing millions towards financing Middle East investments" was correct. These investments enabled the production of dollar oil in the Middle East to grow much more rapidly than non-dollar oil. Between 1947 and 1950 Middle East dollar oil production rose by almost 150 per cent while that of non-dollar oil rose by only 85 per cent.[67] The Marshall Plan funds thus greatly accelerated the process of displacement of the latter (mainly British) oil.

### FINANCIAL ASSISTANCE

The financial help extended to companies did not stop with the Marshall Plan and the other foreign aid programs, and had in some cases begun before the end of the war. Among the other forms of assistance provided by the U.S. government was the decision of the U.S. in 1944 to allow the Saudi Arabian government to purchase gold from the U.S. Treasury at a price of $35 per ounce.[68] This decision greatly helped ARAMCO in a dispute it was having with the Saudis. Under the terms of the original concession agreement, ARAMCO was to pay Saudi Arabia a royalty of four gold shillings per ton of crude or its dollar or sterling equivalent.[69] The company paid royalties and advances to the government in gold sovereigns until the outbreak of World War II, when, because of embargoes, gold became unavailable. During the war it made payments in dollars. A controversy then arose over how gold should be valued in terms of dollars and pounds. ARAMCO argued that the relevant rates were those in New York and London, while the Saudis claimed that the rates in Jiddah should apply. At that time the par value of the English gold pound in London was $8.2397. In Jiddah the English gold pound was selling at between $16 and $20,[70] or approximately twice the London price. Translated into dollar–gold equivalents, the London price was equal to $35 per ounce while the Jiddah price fluctuated around $70 per ounce. By allowing the Saudis to use their dollars, which were paid to them by ARAMCO, to buy gold at $35 an ounce, the U.S. Treasury relieved ARAMCO of paying the expensive Jiddah price for equivalent amounts of gold. It should be remembered that under the U.S. laws existing at that time the Treasury was forbidden to sell gold to individuals and corporations. It could sell gold only to "friendly" governments. Saudi Arabia was obviously considered a "friendly" government.

Another instance of financial help to ARAMCO came in 1949, when the Saudi government was demanding increased payments from the company. The Saudi finance minister suggested that ARAMCO pay the Saudi government an income tax and credit that tax against its U.S. tax obligations.[71] Under the original concession agreement, ARAMCO was exempted from any income tax payments to Saudi Arabia; it was required to pay only a royalty. A royalty, technically, is not a tax but a payment for the right to extract the oil. Under the U.S. income tax laws, royalties were considered an expense and, as such, could not be fully written off against U.S. tax obligations. If, for instance, the corporate income tax rate was 50 per cent and the increase in royalties was $1 million, then the company could reduce its U.S. tax obligations by only $500,000, *i.e.*, by 50 per cent of $1 million. An increase in income taxes of $1 million could, on the other hand, be fully written off against U.S. tax obligations.

The problem for ARAMCO was that, while the U.S. government allowed its companies to write off their foreign income tax payments, these write-offs had to be genuine. Since ARAMCO was exempt from any income tax payments, the Saudi government could not, presumably, impose an income tax on ARAMCO without its consent. If ARAMCO gave its consent, ARAMCO officials feared that the U.S. tax authorities might view this consent as a means to evade U.S. taxes. The authorities could, after all, point out that ARAMCO could increase its royalties to the Saudis to satisfy their demand for more income. In addition ARAMCO was concerned that the U.S. tax authorities might view with suspicion the fact that ARAMCO financial experts had examined Saudi Arabia's finances and were in a position to shape that country's tax structure.[72]

In July 1949 the officials of ARAMCO voiced their concerns to the State Department. They met with a group headed by George McGhee, assistant secretary of state for Near East, South Asian, and African affairs. McGhee was himself an oil man. He was sole owner of McGhee Production Company, an independent oil company.[73] He had previously been treasurer of the Creole Oil Company,[74] a Rockefeller-controlled company in Venezuela whose parent, Standard Oil (New Jersey), was a partner in ARAMCO. Upon leaving the State Department he became a director of Mobil,[75] also a partner in ARAMCO. McGhee was thus in a position to understand fully the problems faced by ARAMCO.

While not committing himself, he promised that "the Department would move as rapidly as possible in the matter."[76] Two days later, a department group met, discussed the matter, and forwarded to McGhee a recommendation "that Aramco should be advised it was the Department's view that the problem was one which primarily concerned Aramco, SAG [Saudi Arabian Government], and the U.S. Treasury." The group further suggested the department informally tell ARAMCO that "if SAG should ask for guidance in tax matters the Department of State would pursue its traditional policy of assisting foreign governments in obtaining competent and independent advisers on tax matters."[77] The substitution of "competent and independent advisers" for ARAMCO's experts presumably would clear ARAMCO of suspicion that it was structuring the Saudi tax laws in such a way as to avoid the payment of U.S. income taxes.

After receiving this recommendation, McGhee first phoned and then despatched a letter to John Graham, assistant secretary of the Treasury, explaining ARAMCO's problem, despite the recommendation of the State Department group that "the problem was one which primarily concerned Aramco, SAG, and the U.S. Treasury."

Soon afterwards the Treasury and ARAMCO sent a team of tax lawyers to Saudi Arabia to draw up a corporate income tax law in a country which had very few corporations and no income tax. They reached an agreement in 1950, under which ARAMCO agreed to submit to a Saudi income tax.[78] In 1951 the partners in ARAMCO began claiming substantial U.S. tax credits for their payments to the Saudi government. Apparently there was a dispute in the Treasury Department over the legitimacy of these tax credit claims. Finally, in 1953 or 1954 the Treasury, in a private ruling not published until 1955,[79] upheld the claims.

The Treasury's actions were most unusual. They caused a substantial loss to the U.S. Treasury. Dr. John Blair, who served as chief economist of the U.S. Senate Subcommittee on Antitrust and Monopoly and was co-author of the U.S. Federal Trade Commission's classic study *The International Petroleum Cartel*, commented:

> Between 1950 and 1951, Aramco's payment to Saudi Arabia rose from $66 million to $110 million, while its payment to the U.S. Treasury fell from $50 million to $6 million. The only loser was the American taxpayer, who had to make up the loss in tax revenues. By 1955, the annual loss to the U.S. Treasury from Aramco had tripled, rising to $154 million. Comparing the company's actual income statement with what would have been the case had it been a domestic U.S. producer, assuming the other entries to have been the same, Aramco, as a domestic concern, would have returned $154,867,000 in taxes to the federal government plus $42,030,000 to state and local governments. Instead $192,743,000 was paid in the form of corporate income taxes to the Saudi Arabian government.[80]

These tax concessions enabled ARAMCO to satisfy the Saudi government's demand for higher revenue, actually 50 per cent of ARAMCO's profits, without costing the company one cent. At the time these concessions were granted, the company was not exactly impoverished. According to Blair, before the imposition of the Saudi income tax the company retained 91 cents a barrel, after paying operating expenses of 20 cents, a royalty to the Saudi government of 21 cents, and a U.S. income tax of 43 cents.[81] In 1950, the last year under this system, ARAMCO produced 200 million barrels of oil,[82] thus receiving, after paying all operating costs, royalties, and taxes, a cash flow[83] of $182 million. Standard Oil (New Jersey) and Socony-Vacuum had paid out $125 million to buy into ARAMCO (which sum had enabled the original partners to recoup their investment). An annual cash flow of $182 million on an investment of $125 million does not imply a low rate of return. Had ARAMCO acceded to the Saudi government's demand through an increase in royalties, rather than through the imposition of an income tax, it would have retained 50 cents a barrel after paying operating costs, a royalty to the Saudis of 88 cents, and a U.S. income tax of 17 cents.[84] On the same production it would have kept $100 million. A cash flow of $100 million on an investment of $125 million still would not have left the company poverty-stricken. As it turned out, after the imposition of the Saudi income tax, ARAMCO retained 91 cents per barrel after paying 21 cents in royalties to the Saudi government, 67 cents in income taxes to that government, and receiving from the U.S. government an "excess tax credit" of 24 cents, which it used to reduce its U.S. tax obligations on income earned in other parts of the world.[85] Its total cash flow on its 1950 production was exactly the same as under the original pure royalty system.

What was most unusual about the Treasury's actions was that they were done in private without the consent of Congress. What, in effect, was a significant appropriation of public funds was performed by a few individuals even though, under the U.S. Constitution, all such appropriations are supposed to be approved by Congress. Surprisingly, this action was never seriously questioned until some twenty years later, when Senator Frank Church, in his capacity as chairman of the Subcommittee on Multinational Corporations, asked George McGhee:

> And I am wondering if the Congress was ever consulted. This would seem to me to be a very ingenious way of transferring many millions by executive decision out of the public treasury and into the hands of a foreign government treasury without ever needing any appropriation or authorization from the Congress of the United States. Isn't that true? Or was Congress asked to authorize this transfer?[86]

All McGhee could say was that "there was, to the best of my knowledge, consultation between the Treasury Department and the appropriate congressional committee."[87] McGhee's apparent lapses of memory were even more interesting. When, for instance, Jack Blum, associate committee counsel, asked him if ARAMCO officials had suggested that "if the U.S. Government wanted more money paid the U.S. Government should pick up some of the tab," McGhee replied that "they didn't go into any detail on a tax advantage with us."[88] Several years after McGhee made this statement, the State Department declassified some hitherto secret material. Below are excerpts from the department's summary of the meeting between McGhee, other State Department personnel, and ARAMCO officials July 20, 1949, to which we already referred.

> Mr. Duce [Vice-President of ARAMCO] said that about a year and a half ago Prince Faisal when he was in America had remarked that Aramco paid a large income tax to the United States Government and should pay some tax to the Saudi Arabian Government. Speaking in the same vein the Saudi Arabian Finance Minister told Mr. Duce last January that the Government of Saudi Arabia needed money. *He said he understood that Aramco could pay Saudi Arabia an income tax and deduct it from the company's U.S. income tax, thus putting no increased burden on Aramco but shifting the flow of dollars from the U.S. Treasury to the Treasury of the Saudi Arabian Government.* (Italics added)

[Mr. Brougham, ARAMCO Vice-President, Financial Matters] went on to say that the Saudis had asked George Eddy, of the Treasury Department, when he was in Saudi Arabia how to get more money out of oil, and Eddy mentioned various methods such as increasing royalties, participating in profits, and the possibility of Saudi Arabia levying an income tax on Aramco. *It seems that Eddy told the Saudis that the sums paid by Aramco to Saudi Arabia in income tax would be deductable* [*sic*] *from U.S. income taxes paid by the company*. (Italics added)

*When Mr. McGhee suggested that the question be put up to the Bureau of Internal Revenue* Mr. Erskine [an attorney for ARAMCO] said the Bureau would probably not give a ruling unless the case was very specific and the terms of the new tax had been worked out in detail. (Italics added)

Mr. Brougham said that the company would like to advise and work with the Saudi Arabian Government through recognized experts as has been done earlier this year when Aramco through Judge Hudson guided Saudi Arabia in formulating a policy for offshore oil. Aramco was afraid, however, that if they did this the Department of Justice would say that the tax was not a genuine one and would disallow Aramco obtaining an exemption on it.[89]

When asked by a committee staff member about how the State Department conveyed its views on the matter to the Treasury, McGhee replied:

The department's foreign policy objectives would have been conveyed through the National Security Council. I assume that the Treasury Department had representatives at whatever large meetings we had at my level in the department. *I don't recall myself ever going to the Treasury Department*.[90] (Italics added)

While there is no evidence that McGhee ever physically went over to the Treasury Department to discuss this problem, he, as has already been noted, both telephoned and wrote the assistant secretary of the Treasury about ARAMCO's dilemma. For some unexplained reason he failed to recall his telephone conversation and letter even though he looked through his records before testifying.[91] Whatever may have been McGhee's reasons in not divulging all the information, he justified the tax decision on the grounds that it was "exceedingly important from the standpoint of the stability of regimes in the area and the security of the Middle East as a whole and the continued ownership of our oil concessions there and the ability to exploit them that the Government of Saudi Arabia receive an increased oil income."[92] Apparently the interests of the United States coincided with those of the feudal Saudi regime and the profits of ARAMCO.

Needless to say ARAMCO was not the only oil company to enjoy the largesse of the U.S. Treasury. Royalties were transformed first into income taxes and subsequently into foreign tax credits in Kuwait, Iran, and other places.[93] There were no apparent limits to the U.S. government's generosity to the oil companies, the flag bearers of the new American Empire.

In addition to the aid given the companies through the Marshall Plan, the sale of gold, and the foreign income tax credit, the U.S. helped them through the exertion of diplomatic pressure against Britain. In 1950 Britain, in an effort to reduce her dollar deficit, took measures to favour the importation of "sterling" oil, *i.e.*, oil produced by Anglo-Iranian and Shell, over "dollar" oil.[94] The State Department, expressing its "emphatic concern," immediately protested.[95] After several months of negotiations, the United States and the United Kingdom reached a compromise under which Britain agreed to ease its restrictions and the U.S. oil companies consented to accept partial payment in sterling.[96]

### IRAN

The U.S. government also played a direct role in opening up Iran to American oil companies. Until 1951, Iran was the private preserve of Anglo-Iranian. The role of the United States in overthrowing Mossadegh in 1953 and returning the Shah has been well documented and has received relatively widespread publicity. The U.S. moves in Iran have generally been interpreted by defenders and critics alike as part of the U.S. Cold War strategy. The general presumption is that the U.S. feared Mossadegh would ally himself with the Soviet Union, and therefore the U.S. moved to overthrow him and replace him with the more reliable Shah.[97] While this is unquestionably true, it is only part of the story. What is generally not so well recognized is that the U.S. action was directed also against Britain. It was the means by which the U.S. was able to break the British oil monopoly in that country.

When Mossadegh nationalized the properties of Anglo-Iranian in 1951, the United States adopted an ambiguous attitude towards this move. On the one hand, as the nation with the largest stake in overseas investment, it expressed concern about the expropriation. John Loftus, head of the State Department's petroleum division, complained:

> While recognizing the sovereign right of any country to assume ownership...of the petroleum industry or any of its branches, this Government must nevertheless recognize and proclaim that international commerce, predicated upon free trade and private enterprise (which is the conceptual core of United States economic foreign policy), is, in the long run, incom-

patible with an extensive spread of state ownership and operation of commercial properties.[98]

On the other hand, the State Department had no intention of using its influence to restore the seized properties to Anglo-Iranian. It, for instance, vetoed an attempt by the British to take military action against Iran.[99] Unquestionably the department feared that such a move might provoke Soviet intervention. But it also did not want the British to regain their former position in the country. As Engler has noted, the nationalization presented the U.S. with "a new opportunity"[100] to bring the "open door" policy to Iran.

The overthrow of Mossadegh, which was engineered by the CIA, paved the way for the displacement of Britain by the United States as the major power in Iran. The displacement took place first in oil and then in the other sectors of the economy.

After the coup the U.S. sent Herbert Hoover, Jr., a director of Union Oil,[101] to negotiate a new oil pact. Under the agreement the properties formerly belonging to Anglo-Iranian were turned over to the National Iranian Oil Company (NIOC), a company wholly owned by the Iranian government. However, oil exploration and production in southern Iran and the operation of the Abadan refinery, then the world's largest, were to be carried out by a consortium of companies[102] known as Iranian Oil Participants Ltd., under contract to the NIOC.[103] As originally envisaged, membership in the consortium was to be restricted to the eight major international oil companies, the cartel companies,[104] namely: Anglo-Iranian, whose name had by then been changed to British Petroleum (BP); Gulf; Socony; Standard Oil (New Jersey); Standard Oil of California; Shell; Texaco; and CFP. BP was to receive 40 per cent; Shell, 14 per cent; CFP, 6 per cent; and the other five companies, all American, were to receive 8 per cent each. Thus, on a national basis the shares of the consortium were distributed as follows: England 46 per cent, United States 40 per cent, Holland 8 per cent, and France 6 per cent.[105]

There was, however, one important problem with the share going to the five U.S. companies. All these companies at that time were undergoing prosecution for violation of the U.S. antitrust laws in what has been referred to as the "Cartel Case."[106] This case, initiated by the Department of Justice in 1952, was partially based on the findings of the Federal Trade Commission's famous report *The International Petroleum Cartel*. Among the stated objectives of the department was the ending of the "monopolistic control of foreign production and of the policy of excluding American independents from foreign sources of supply."[107] It is ironic, to say the least, that at the very time the Justice Department was pressing this suit, the State Department negotiated a treaty which gave the American cartel companies a greater control of foreign production than ever before and which excluded American independent companies from Iran.

However, this incongruity did not deter U.S. policy-makers. The Justice Department merely issued an advisory opinion that the agreement did not violate the antitrust laws.[108] To make their highly imaginative interpretation of the law a little more palatable, the department's officials persuaded each of the five U.S. companies to surrender a 1 per cent share to American independent companies. Thus the independents, as a group, were to receive a 5 per cent share, and each of the five U.S. internationals was to receive 7 per cent.[109]

Once this decision was made the question arose as to how to select the independent companies to participate in the consortium. Since the State Department did not want to be put in the position of favouring one U.S. firm over another, it selected the accounting firm of Price Waterhouse and Company to determine the eligibility of the applicant firms. That company undoubtedly had the qualifications to perform the screening process: among its clients were Standard Oil (New Jersey), Standard Oil of California, Gulf, and Shell.[110] The choice of this company evoked a bitter protest from Watson Snyder, the petroleum specialist in the Justice Department's Antitrust Division. Snyder complained that "all through the documentary material delivered by the five defendants in the cartel case, you will find that Price Waterhouse and Company is the medium through which all the accounting is done for the participants in the various illegal arrangements. . . . Whenever either the domestic or foreign branches of the petroleum industry carry out any joint operations Price Waterhouse is chosen to do the accounting."[111] As Blair pointed out, "Price Waterhouse was not likely to certify any firm strongly objected to by the majors."[112] This incident illustrates both the intra-industry problems generated by the policy of supporting private American firms in their quest for overseas reserves and the close links between the State Department and the large international oil companies.

Price Waterhouse performed its duty faithfully. It approved every U.S. applicant except one, the International Cooperative Petroleum Association, whom it vetoed on the grounds that the Association was not truly an American company "in that it has members in other countries." There is, however, evidence that the international oil companies, which were consulted by Price Waterhouse, feared that the Association's "participation in the sale of Iranian oil in Western Europe could upset prevailing price structures."[113] Originally Price Waterhouse approved twelve U.S. firms, who together formed Iricon Agency Limited, holding five per cent of the shares in the consortium. Mergers and sales of shares reduced the number of participants in Iricon. In 1974 they were six: Atlantic Richfield, which owned one-third of Iricon's shares; American Independent, Getty, and Charter, each of which owned one-sixth; and Sohio and Conoco, each of which owned one-twelfth.[114]

The exemption of the consortium members from antitrust action spelled the doom of the "Cartel Case" against the five U.S. internationals. The case originally covered all phases of the industry's entire international operations. But the charges relating to monopolistic practices in production, transportation, storage, and refining were dropped on the grounds that the prevailing practices were less restrictive than those countenanced in the consortium agreement. Hence the exemption of that agreement implied the exemption of the other practices. The Antitrust Division dismissed all charges without prejudice against Mobil and Standard Oil of California. The whole case ended in 1968 without any substantial impact on the industry.[115]

There were several important repercussions following the establishment of the consortium, however. First, it set up for the first time a joint venture in which all the eight cartel companies participated. Because of this particular characteristic of the consortium, Iran played the pivotal role in determining world production. Once Iranian production was set, all the partners in the consortium set their production in other parts of the world accordingly. The companies used a formula, Average Program Quantity (APQ), through which they coordinated Iranian production with production in the other countries in which they had concessions.[116] This was a cartel *par excellence*.

Second, though BP lost its monopoly in Iran, it did not lose financially. The consortium members estimated the value of BP's assets at $1 billion. Accordingly, the other participants paid BP $600 million for their 60 per cent share in the consortium. The Iranian government, under the newly restored Shah, also agreed to pay BP $70 million as indemnity.[117] This indemnity was a major factor in inflaming Iranian opinion against the consortium and against the Shah. There was popular feeling that BP had taken far more than its just share of profits long before the nationalization; there was no justification to pay it any indemnity. By accepting this indemnity, BP, along with the other members of the consortium, helped sow the seeds of the Islamic Revolution which was to sweep the country some twenty-five years later.

Third, the establishment of the consortium was the most important factor in making the U.S. the dominant oil power in the Middle East. The entry into Iran unquestionably gave it effective control of most of the known reserves in the non-Communist world. Its basic objective, the control of world oil, had been realized. This was to have important implications, not only in the Middle East, but throughout the world.

**Notes**

1. N. Jacoby, *Multinational Oil, A Study in Industrial Dynamics* (Macmillan, New York, 1974), p.37.
2. U.S. Senate, Subcommittee on Multinational Corporations of the Committee on Foreign Relations, *A Documentary History of the Petroleum Reserves Corporation, 1943–1944*, 93rd Congress, 2nd Session (U.S. Government Printing Office, Washington, 1974), p. 3.
3. *Historical Statistics of the United States, Colonial Times to 1957: A Statistical Abstract Supplement* (U.S. Government Printing Office, Washington, 1960), p. 143.
4. Ibid., p. 735.
5. In only one other conflict, the U.S. Civil War, did the U.S. forces fight for as long a time as in World War II. In the Civil War the ratios of total casualties and deaths to the armed forces personnel were considerably higher, about 29 per cent for casualties and 17 per cent for deaths. Ibid.
6. Among the works dealing with this aspect of the origins of the Cold War are G. Alperovitz, *Atomic Diplomacy: Hiroshima and Potsdam, the Use of the Atomic Bomb and the American Confrontation with Soviet Power* (Simon & Schuster, New York, 1965); D. Fleming, *The Cold War and Its Origins 1917–1960* (Allen & Unwin, London, 1961); D. Horowitz, *The Free World Colossus: A Critique of American Foreign Policy in the Cold War* (Hill & Wang, New York, 1965); G. Kolko, *The Politics of War: the World and United States Foreign Policy 1943–5* (Vintage Books, New York, 1968); G. Kolko, *The Roots of American Foreign Policy: An Analysis of Power and Purpose* (Beacon Press, Boston, 1969); R. Maddox, *The New Left and the Origins of the Cold War* (Princeton University Press, Princeton, N.J., 1973); J. Morray, *From Yalta to Disarmament, Cold War Debate* (MR Press, New York, 1961); and W. A. Williams, *The Tragedy of American Diplomacy* (Dell Press, New York, 1962).
7. H. Magdoff, *The Age of Imperialism, The Economics of U.S. Foreign Policy* (Modern Reader Paperbacks, New York and London, 1969), p. 42.
8. P. B. Kenen, *Giant Among Nations, Problems in United States Foreign Economic Policy* (Rand McNally, Chicago, 1960), p. 159.
9. H. Magdoff, *The Age of Imperialism*, p. 126.
10. U.S. Senate, Committee on Foreign Relations, *Hearings on Foreign Assistance of 1962* (U.S. Government Printing Office, Washington, 1962), p. 27.

11. U.S. Senate, Committee on Foreign Relations, *Some Important Issues in Foreign Aid* (U.S. Government Printing Office, Washington, 1966), p. 15.
12. The trade balance is the difference between *merchandise* exports and *merchandise* imports. As such, it should be distinguished from other balances, such as the balance of payments, which encompasses all international transactions including service and capital flows as well as merchandise movements.
13. The balance on goods and services encompasses the trade balance and the balance on services like insurance, travel, freight and investment income.
14. *Economic Report of the President Transmitted to the Congress February 1974* (U.S. Government Printing Office, Washington, 1974), p. 352.
15. *Economic Report of the President Transmitted to the Congress January 1980* (U.S. Government Printing Office, Washington, 1980), p. 318.
16. Ibid.
17. M. Wilkins, *The Maturing of Multinational Enterprise: American Business Abroad from 1914 to 1970* (Harvard University Press, Cambridge, Mass., 1974), p. 329.
18. H. Aitken, *American Capital and Canadian Resources* (Harvard University Press, Cambridge, Mass., 1961), pp. 66–67.
19. U.S. Senate, *A Documentary History of the Petroleum Reserves Corporation*, p. 3.
20. Ibid.
21. Ibid., p. 4.
22. Ibid.
23. Ibid., p. 5.
24. Ibid.
25. Ibid., p. 6.
26. Ibid., p. 16.
27. Ibid., p. 35.
28. Ibid., p. v.
29. Ibid., p. 61.
30. Ibid., p. 63.
31. Ibid.
32. Ibid., p. 65.
33. Ibid., p. 63.
34. Ibid.
35. W. J. Kennedy (ed.), *Secret History of the Oil Companies in the Middle East* (2 vols., Documentary Publications, Salisbury, N.C., 1979), vol. 1, p. 104.
36. Ibid., p. 105.
37. Ibid., p. 104.
38. Ibid., p. 8.
39. Ibid., p. 17.
40. Jacoby, *Multinational Oil*, p. 39.
41. U.S. Federal Trade Commission, *The International Petroleum Cartel* (U.S. Government Printing Office, Washington, 1952), p. 120.
42. C. Issawi & M. Yeganeh, *The Economics of Middle Eastern Oil* (Praeger, New York, 1962), p. 180.

43. Kennedy, *Secret History of the Oil Companies*, vol. 1, pp. 75–76.
44. U.S. Federal Trade Commission, *The International Petroleum Cartel*, p. 103.
45. Ibid., p. 104.
46. Ibid.
47. Ibid., pp. 103–5.
48. Kennedy, *Secret History of the Oil Companies*, vol. 1, p. 77.
49. E. Groen, "The Significance of the Marshall Plan for the Petroleum Industry in Europe—Historical Review of the Period 1947–1950" in U.S. Congress, *The Third World Petroleum Congress* (U.S. Government Printing Office, Washington, 1952), p. 49.
50. J. Kolko & G. Kolko, *The Limits of Power, The World and United States Foreign Policy, 1945–1954* (Harper and Row, New York, 1972), p. 447.
51. J. Stork, *Middle East Oil and the Energy Crisis* (Monthly Review Press, New York and London, 1975), p. 61.
52. P. Odell, *Oil and World Power: Background to the Oil Crisis*, 3rd ed. (Penguin Books, Middlesex, Eng., 1974), p. 99.
53. U.S. House of Representatives, Select Committee on Small Business, *Hearings, Effects of Foreign Oil Imports on Domestic Producers*, 81st Congress, 1st Session (U.S. Government Printing Office, Washington, 1949), part 2, p. 523.
54. Ibid., p. 522.
55. Ibid., p. 531.
56. The ECA did eventually finance some expansion in refining capacity, including capacity by foreign companies. It did so because, in the words of Bransky, any cut-off of funds probably would have resulted in "a drastic limitation of dollar oil imports resulting in an even greater increase in expansion of nondollar projects." Though the European countries wanted to increase capacity from 480,000 barrels per day to well over one million barrels, the ECA refused to fund any expansion above 840,000 barrels. In addition the ECA refused to finance any government-owned refineries. See ibid., p. 532.
57. Ibid., p. 530.
58. Ibid., pp. 530–31.
59. E. Shaffer, *The Oil Import Program of the United States: An Evaluation* (Praeger, New York, 1968), p. 13.
60. Kennedy *Secret History of the Oil Companies*, vol. 2, p. 245.
61. H. Frank, "The Pricing of Middle East Crude Oil," unpublished Ph.D. thesis, Columbia University, 1961, p. 58.
62. R. Engler, *The Politics of Oil, A Study of Private Power and Democratic Directions* (The Macmillan Company, New York, 1961), p. 220.
63. Ibid., pp. 522–23.
64. Ibid., p. 523.
65. Frank, "The Pricing of Middle East Oil," p. 86.
66. Ibid.
67. E. Shaffer, "The Oil Import Program of the United States: An Evaluation," unpublished Ph.D. thesis, Columbia University, 1966, p. 65.
68. Kennedy, *Secret History of the Oil Companies*, vol. 1, p. 86.

69. Ibid., pp. 81–82.
70. Ibid., p. 81.
71. Ibid., vol. 2, p. 425.
72. Ibid., pp. 425–26.
73. U.S. Senate, Subcommittee on Multinational Corporations of the Committee on Foreign Relations, *Hearings, Multinational Corporations and United States Foreign Policy*, 93rd Congress, 2nd Session (U.S. Government Printing Office, Washington, 1974), part 4, p. 83.
74. Kennedy, *Secret History of the Oil Companies*, vol. 2, p. 428.
75. L. Turner, *Oil Companies in the International System* (The Royal Institute of International Affairs, London, 1978), p. 104.
76. Kennedy, *Secret History of the Oil Companies*, vol. 2, p. 429.
77. Ibid., p. 432.
78. J. Blair, *The Control of Oil* (Pantheon, New york, 1976), p. 199.
79. U.S. Senate, *Multinational Corporations and United States Foreign Policy*, part 4, p. 124.
80. Blair, *The Control of Oil*, pp. 200 and 202.
81. Ibid., p. 201.
82. Issawi & Yeganeh, *The Economics of Middle Eastern Oil*, p. 183.
83. The cash flow of a company is equal to after-tax profits plus capital consumption allowances. The cash flow is what is left, after paying all current expenses and taxes, either for distribution to shareholders as dividends or for investment. In their accounting oil companies used the "discounted cash flow," rather than profits *per se*, to assess the viability of any given investment. See M. Tanzer, *The Political Economy of International Oil and the Underdeveloped Countries* (Beacon Press, Boston, 1969), pp. 32–33.
84. Blair, *The Control of Oil*, p. 201.
85. Ibid., pp. 200–201.
86. U.S. Senate, *Multinational Corporations and United States Foreign Policy*, part 4, p. 89.
87. Ibid.
88. Ibid., p. 92.
89. Kennedy, *Secret History of the Oil Companies*, vol. 2, pp. 425–27.
90. U.S. Senate, *Multinational Corporations and United States Foreign Policy*, part 4, p. 90.
91. Ibid., p. 88.
92. Ibid., p. 89.
93. Blair, *The Control of Oil*, p. 199.
94. H. Menderhausen, *Dollar Shortage and Oil Surplus in 1949–1950* (Princeton University Press, Princeton, 1950), p. 2.
95. Ibid., p. 11.
96. Ibid., pp. 28–32.
97. See, for instance, F. Halliday, *Iran Dictatorship and Development* (Penguin Books, Middlesex, 1979), p. 25, and Stork, *Middle East Oil and the Energy Crisis*, p. 52.
98. Stork, *Middle East Oil and the Energy Crisis*, pp. 51–52.
99. Ibid., p. 50.

100. Engler, *The Politics of Oil*, pp. 202–3.
101. Turner, *Oil Companies in the International System*, p. 104.
102. Jacoby, *Multinational Oil*, p. 97.
103. Issawi & Yeganeh, *The Economics of Middle Eastern Oil*, p. 175.
104. See U.S. Federal Trade Commission, *The International Petroleum Cartel*.
105. J. Walden, "The International Petroleum Cartel in Iran—Private Power and the Public Interest," *Journal of Public Law*, vol. 2, no. 1 (1962), p. 49.
106. *United States* v. *Standard Oil Company (N.J.)* 86 Civ. 27, (S.D.N.Y.).
107. Blair, *The Control of Oil*, p. 72.
108. Walden, "The International Petroleum Cartel in Iran," p. 48.
109. Ibid., p. 150.
110. H. O'Connor, *The Empire of Oil* (Monthly Review Press, New York, 1962), p. 330.
111. Blair, *The Control of Oil*, p. 45.
112. Ibid.
113. Walden, "The International Petroleum Cartel in Iran," p. 50.
114. Blair, *The Control of Oil*, p. 46.
115. U.S. Senate, *Multinational Corporations and United States Foreign Policy*, part 4, p. 189.
116. Blair, *The Control of Oil*, pp. 101–8.
117. Walden, "The International Petroleum Cartel in Iran," p. 52.

CHAPTER SIX

# The Control of World Oil: Policy and Problems

That the very success of U.S. oil policy in Iran carried the seeds of its own destruction was not apparent to U.S. policy-makers until some two decades later. More immediately apparent were the interrelated problems of State Department relations with individual oil companies and the management of the oil surplus that was then engulfing the world.

## Relationship with Individual Companies

Iran illustrated the dilemma the U.S. government faced in implementing its policy of encouraging U.S. nationals to find reserves overseas. In a country with many oil companies, such a policy must inevitably favour some companies over others. Logically the national interest would lie in having a few firms engage in foreign oil operations. It would then be relatively easy for the State Department to monitor their activities and to coordinate them with the department's policies. The entry of a large number of firms into overseas fields would render such monitoring and coordination difficult.

However, the logic of foreign policy conflicted with the dogma of American free enterprise. This dogma, as enshrined in the antitrust laws, dictated that the government should play no favourites among companies, except, perhaps, to aid small firms in their competitive battles with large ones.[1] The exigencies of foreign policy, on the other hand, demanded that government should deal chiefly with those enterprises experienced in foreign operations, *i.e.*, with the large international oil companies.

This conflict was exemplified by the original attempt of the State Department to invite only the five large U.S. internationals to participate in the Iranian consortium. Pressure from other oil companies and from the Antitrust Division of the U.S. Department of Justice forced the

State Department to allow some smaller American companies to buy into the consortium. The State Department, as has been pointed out, appointed an accounting firm to screen the applicants in order to preserve its official position of maintaining neutrality among competing American firms. In reality the appointment of a company with close links to the internationals indicated the department's bias towards these firms.

### *Control of Entry*

Because diplomatic protection was an essential ingredient of most foreign investment decisions, the State Department was in a position to control entry into foreign oil. Firms entering an area without the department's blessing did so at their own risk. Generally the department used its power to protect the internationals against potential competitors. This screening process was most evident in areas of the Middle East. An example can be seen in the treatment of Superior Oil in Saudi Arabia.

#### SAUDI ARABIA

In July 1947 three executives of Superior Oil arrived in Jiddah, Saudi Arabia, seeking to meet government officials to discuss possible concession terms. Though they had telegraphed in advance to the U.S. minister, J. Rives Childs, the time of their arrival, the minister chose not to meet them at the airport, sending instead his third secretary. When William Keck, the executive officer of Superior, asked Childs to arrange an appointment to meet King Ibn Saud, Childs refused to do so. In a despatch to the State Department Childs wrote that "I informed Keck that it was unnecessary; that he need have no fear but that the King was aware of the presence of the Superior Oil Company group in Saudi Arabia."[2] Despite Childs's misgivings about the Superior contingent,[3] he arranged for them to meet several local Saudi officials. He also entertained them for dinner, inviting, among others, the local representative of ARAMCO. In his report to the State Department Childs explained:

> I thought it important to include Mr. Campbell of Aramco, in order not to give the impression to the Saudi Arabian Government that we were playing up the Superior Oil Company in the midst of Aramco's difficulties in a way which might confuse the Saudi Arabian authorities. I believe Mr. Campbell appreciated the care I took in that regard."[4]

The problems to which Childs referred related to the location of the proposed pipeline to the Mediterranean, the financing and operations of railroads and piers, the gold-pound controversy, and the Saudi demand for increased royalty payments. Of particular concern to both ARAMCO and the U.S. government at that time was the location of the terminus of the pipeline. The construction of the line was considered a top priority to provide outlets both to satisfy King Ibn Saud's demands for increased production and to replace Western Hemisphere oil in European markets with oil from the Middle East. Pipeline construction was being delayed because of real estate speculation by Saudi officials. A secret State Department memo of June 6, 1947, for instance, revealed:

> The exact location of the terminus of the Trans-Arabian pipeline on the eastern shore of the Mediterranean has been complicated by the fact that various Saudi Arabian officials have been speculating in land at the places where they thought the line would end.[5]

Needless to say, the necessity of mollifying the conflicting Saudi interests involved ARAMCO in delicate negotiations with various layers of Saudi officialdom. The entrance of an outsider, who might offer these officials a different set of proposals, could obviously upset these negotiations. Such a development would cause ARAMCO justified concern. Whether it should also cause concern to the U.S. government is, of course, a matter of contention.

The officials of Superior, quite understandably, did not sympathize with the concern expressed by the U.S. minister. In a meeting with State Department officials on May 6, 1948, the company's vice-president, Carlton Wood, voiced dissatisfaction with Childs's treatment of Superior's personnel. He complained specifically that Childs would not grant the company's representatives an interview, refused to house them in the legation, and discussed Superior's plans with ARAMCO officials. Loy Henderson, the department's director of Near Eastern and African Affairs, assured Wood that "Mr. Childs intended no such discrimination but was probably quite apprehensive lest any action on his part be construed as indicating official U.S. Government sympathy to one party or the other."[6]

What Henderson neglected to mention was that the department apparently was not concerned about giving the appearance of favouritism when it dealt with ARAMCO. When Childs, for instance, notified the department on March 20, 1947, just two months before the arrival of the Superior group, that "Mr. D. C. Dennett, of the American legation in Beirut, arrived in Jidda on March 14, 1947, *in an Aramco plane from Dhahran on a brief tour of the Middle East during which he traveled with*

*Mr. William Lenahan of the Arabian American Oil Company*"[7] (italics added), he expressed no misgivings about the impression such a travel arrangement might have on the Saudi government. Nor apparently did the State Department.

The department still continued to profess its neutrality, which was frequently questioned by Superior. In a meeting with department officials on June 17, 1948, Robert Allen, Jr., Superior's vice-president in charge of foreign operations, complained that ARAMCO officials knew about his company's discussions with the Saudi government. He wanted to know whether the department was really neutral in its treatment of ARAMCO and Superior.

According to the department's memorandum of the meeting, Henderson told him that

> the Department continued to favor competition and the appearance of new companies as concession holders in the Middle East but could not to this end favor any one company over another. The Department has endeavored to treat all companies alike and felt that this position must be maintained.[8]

This type of neutrality, to the extent that it existed, was, of course, no neutrality at all. It favoured the powerful, established firms over the newcomers. It is reminiscent of Anatole France's famous dictum, "the law in its majestic equality forbids the rich as well as the poor to sleep under the bridges, to beg in the streets and to steal bread." But, despite its professions to the contrary, the State Department did not practise even this type of neutrality. It took positive action on behalf of ARAMCO.

Superior and ARAMCO were involved in a dispute over the right to develop Saudi Arabia's offshore oil resources. Superior had been seeking offshore concession rights from the Saudi government. The State Department was opposed to the granting of these rights until an agreement could be reached with Britain and the states bordering the Persian Gulf on how to apportion the sub-surface rights among the interested parties. The ceding of these rights by Saudi Arabia to Superior or any other oil company would jeopardize the negotiations leading to such an agreement. In addition, there was the question as to whether ARAMCO's concession would apply to those areas granted to Saudi Arabia under an international agreement. The department's summary of the conversation with Superior's Wood on this latter question reads as follows:

> Mr. Wood. . .emphasized that he did not feel that this second problem was one which involved the Department of State. Mr. Henderson agreed that this was so as regarded the matter of interpretation of the Aramco concession,

but he added that it would be a most unfortunate development should a major dispute arise between American companies interested in the Persian Gulf. Such a dispute would be embarrassing to the United States in view of various political factors affecting the entire Middle East.[9]

How competition between two American companies for concessions in the same region could be "embarrassing to the United States," the world's foremost advocate of a competitive economy, Mr. Henderson did not say. In any event the State department avoided this "embarrassment" by giving informal backing to ARAMCO.

In a meeting on June 16, 1948, one day before Henderson assured Superior officials that the "Department has endeavored to treat all companies alike," U.S. Minister Childs had a meeting with Sheikh Yusuf Yassin, deputy foreign minister of Saudi Arabia, in which they discussed the ARAMCO-Superior dispute. The Saudi government had received a favourable offer from Superior for offshore concessions. It included payment to the government of a minimum of 50 per cent of net profits, rising on a sliding scale to 75 per cent, as well as the appointment of a Saudi general manager, who would have equal status and salary with the foreign general manager. As Childs pointed out, the appointment of a Saudi general manager "has made a very special appeal to the Saudi Arabian Government in view of the strong feeling that has existed for some time in the minds of Saudi Arabians against what they regard as discrimination on the part of Aramco against Saudi Arabians in the provision of living accommodations and social and economic equality of treatment."[10] Upon receiving this offer, the Saudi government had given ARAMCO until June 30, 1948 to make a counter offer. In reply ARAMCO requested a delay until July 31.

Sheikh Yusuf asked Childs for advice on how to handle the matter. In a secret despatch to the State Department, Childs recorded his reply:

I stated to Shaikh Yusuf that I felt this to be a matter between the Saudi Arabian Government and Aramco with which the United States was not officially concerned. He pressed me for my personal view.

I stated that speaking entirely personally I would answer him on the basis of a homely analogy. If I had been doing business with a firm or individual over a period of some years, and that firm or individual asked me for time in which to consider a problem arising through the interposition of a third, and hitherto unknown individual or firm, I would feel that as a matter of courtesy, if nothing else, the individual or firm with whom I had been doing business was entitled to obtain the delay desired, particularly if no prejudice was done thereby to either party.

Shaikh Yusuf nodded and appeared to accept my reply to his question as a reasonable one.[11]

Perhaps it is not surprising that Childs, who had previously praised ARAMCO for its "very high sense of social responsibility,"[12] should on his own initiative violate the proclaimed U.S. policy of maintaining neutrality between competing U.S. companies. What is surprising, however, is that he felt secure in reporting this violation to his superiors. What is even more surprising is that his superiors took no action against him.

Whatever were the real reasons for ARAMCO's request for a delay is a matter of speculation. Any delay, for whatever reason, was bound to work to ARAMCO's advantage. Superior did not have the financial resources of ARAMCO. It could not afford to wait as long as ARAMCO for a return on its investments. Unless Childs was exceptionally naive, he must have known that any delay would have had a much greater financial impact on Superior than on ARAMCO. By suggesting that the Saudis accede to ARAMCO's request, he was in effect helping ARAMCO raise the cost of entry for Superior.

His superiors also helped delay a final decision. On July 21, 1948, three executives of Superior held further discussions with State Department officials in Washington. The Superior group claimed that the Saudis were ready to grant Superior the concession provided that the U.S. government assured the Saudis that the area came under Saudi jurisdiction and that the U.S. government would not interfere with the granting of a concession to a firm other than ARAMCO.

The department refused to grant these assurances on the grounds that "it would be unwise and dangerous for any concession to be granted covering the submerged area of the Persian Gulf until some orderly division of the submerged area as between the riparian states had been effected."[13] When Wood, of Superior, pointed out that the concession was in an area not subject to conflicting claims from bordering states and therefore there was no reason why the granting of the concession should be delayed, Satterthwaite, of the State Department, insisted otherwise. According to the department's memorandum of the meeting, Satterthwaite stated:

> . . . the Department by all means desired that the final orderly division of the submerged area be awaited before any concession be granted. It was especially important to do this because if individual rulers were to start issuing concessions before their title to do so had been made clear the way would be left open for the Russians to enter the Gulf and start drilling anywhere they chose.[14]

When Allen, of Superior, noted that the Soviets could be excluded by either a unilateral declaration of sovereignty over a concession area or an agreement among the states, Satterthwaite replied that an agreement "was preferable in that it would provide a more orderly approach to a peaceful solution of the problem."[15]

The department's refusal to endorse Superior's candidacy in any way spelled the end of Superior's chances of getting into Saudi Arabia. Fearful of alienating the U.S., the Saudi Arabian government allowed the negotiations to drag on for almost a year, until ARAMCO sent a delegation to Saudi Arabia with the text of a proclamation on offshore mineral rights. This proclamation was adopted unilaterally by the Saudi government on May 29, 1949.[16] Needless to say, ARAMCO retained its rights to the offshore areas. ARAMCO's actions evoked no apparent protest from the State Department. When ARAMCO executives met with State Department officials on June 17, 1949 to report on their role in the Saudi proclamation, no one in the department seemed at all concerned with the repercussions of ARAMCO's behaviour on other states and oil companies in the region. Whether by accident or design, the State Department's actions effectively blocked the entry of Superior into Saudi Arabia. The "open door" policy apparently did not extend to those American companies which did not meet with the department's approval.

### AFGHANISTAN

In addition to regulating entry, the State Department attempted to steer companies to certain specific areas. In January 1948 Charles Rayner, Washington representative of the American Independent Oil Company, a firm owned by approximately twelve independent U.S. companies and controlled by Phillips Petroleum and Signal,[17] visited Loy Henderson at the State Department to discuss Rayner's proposed trip to the Middle East. Rayner had previously been the State Department's petroleum adviser. Rayner inquired about the feasibility of going to Iran, which was then the exclusive domain of Anglo-Iranian. Henderson, according to the department's memorandum of the meeting, told him that "such a trip by an American oil man did not now seem desirable."[18] Henderson presumably feared that such a trip might upset the negotiations with the British over riparian rights.

Henderson did, however, suggest to Rayner an interesting alternative, Afghanistan. In view of events in that country some thirty years later, Rayner's reply turned out to be quite revealing. According to the State Department, Rayner felt

> that country was too inaccessible to justify exports and that the domestic market in Afghanistan was at present very restricted. Furthermore the local

terrain was difficult and the proximity of the Russians was a constant deterrent to American investment in Afghanistan.[19]

One can only speculate whether some thirty years later, when the United States was desperately trying to find new sources of oil to replace its losses in Iran, it attempted to open up Afghanistan to American investment through reducing Soviet influence in that country, and if the Soviets reacted to this attempt by intervening in Afghanistan. Whatever the truth may be, it is certainly worth noting that the United States has been interested in exploring in Afghanistan since the end of World War II.

Henderson's suggestion indicates how the State Department tried to direct the smaller U.S. companies to specific, and often less desirable, areas. While this direction may have been perfectly rational from the point of view of assuring entry by American companies in all parts of the world, it had the effect of protecting the international oil companies from competition from newcomers in the areas in which the internationals had previously gained a dominant position. The department's role was that of a cartel "secretariat," controlling entry so as not to disturb the equilibrium reached by the cartel.

*Growth of Foreign Reserves*

However, the State Department was not able to exercise complete control over entry. Because of specific economic conditions in the U.S. many smaller firms sought to obtain foreign oil reserves on their own. In addition, the tax legislation existing at that time gave them an incentive to invest abroad.

In the years following World War II the domestic industry reaped enormous profits, for the rapid expansion of the economy benefited the oil industry especially. This expansion was both motivated by and resulted in a large increase in motor car ownership. Between 1946 and 1953, for instance, automobile registrations in the U.S. rose from 34.4 million to 56.3 million,[20] an increase of 64 per cent. Gasoline consumption consequently soared from 30 trillion gallons per year to 49 trillion,[21] an annual rate increase of more than 7 per cent.

In response to this increasing demand, petroleum prices climbed much more sharply than prices of other commodities. The wholesale price of petroleum and its products advanced by 70 per cent while those of all other commodities rose by only 40 per cent.[22] As a result, the flow of profits to the refining industry, as reported by the Federal Trade Commission and the Securities and Exchange Commission, was higher than in any other manufacturing industry in every year from 1947 to

1953. In these seven years the refining industry amassed $13 billion in after-tax net profits; its profits accounted for one-sixth of the total profits of all manufacturing industries.[23] The depletion allowance provision of the income tax, which allowed firms producing crude oil to deduct 27½ per cent of the value of their crude oil sales from their taxable income, further enhanced the flow of funds to the industry.

Initially, a sizable portion of these funds was invested in the domestic industry. Proved reserves consequently rose by 38 per cent between 1946 and 1953.[24] Production capacity, *i.e.*, the capacity to produce oil from existing wells, jumped by approximately 50 per cent.[25] However, the quantity demanded (as measured by domestic production plus imports plus changes in stocks) increased at a much slower rate, by only 33 per cent.[26] This relatively slow rate of growth in demand placed downward pressure on domestic oil prices.

## DOMESTIC PRODUCTION CONTROLS

The authorities in Texas and other oil-producing states countered this downward pressure by restricting production. Since the 1930s the U.S. government has permitted the oil-producing states to restrict production through the imposition of "market demand prorationing." Officially justified as a "conservation" measure, this type of prorationing turned out to be a price maintenance system. As practised in Texas and most of the other oil states, the system empowered a regulatory authority to place a ceiling on the amount of oil that could be produced each month. This ceiling was determined essentially by the "nominations" made by refiners to the regulatory agency. For instance, the Texas Railroad Commission, that state's regulatory authority, required all refiners intending to purchase oil in the state to notify it about the quantities they wanted to purchase. The commission then restricted output to these quantities.

These restrictions prevented a "surplus" of oil from being produced, "surplus" being defined as the production of more than is demanded at the prevailing price. The prevention of this "surplus" had the intended effect of either reducing or completely eliminating downward pressure on prices. In order to prevent total production from exceeding refiners' "nominations," the regulatory agency assigned each well a production quota, called an "allowable." Any well-owner who exceeded his "allowable" faced severe economic and legal sanctions. The prorationing system thus cartelized the U.S. domestic oil industry, with the state regulatory bodies playing the role of a cartel "secretariat" through their setting of production ceilings and their assignment of "allowables."

During the 1950s the state authorities started cutting back "allowables" significantly. While these cutbacks maintained or even increased crude prices, they began to threaten the profit potential of many wells since producers could not sell as much oil as before. This reduction in profit potential led many oil companies to look for reserves abroad, where, with the exception of Canada, prorationing did not exist. Thus, instead of investing their profits in the United States, they sent them outside the country in an effort to keep their future profits from falling.

The depletion allowance gave integrated refining companies (those companies owning both refineries and producing wells) a further incentive to invest abroad. The reductions of "allowables" meant that many refining companies were not permitted to produce from their own wells as much oil as they required, but were forced to buy oil from other producers. When they did this, they lost the benefit of the depletion allowance, which could be claimed only by sellers of crude oil, at the same time that they gave this advantage to their competitors. By producing abroad they could take full advantage of the depletion allowance. In addition they could, as has already been pointed out (see page 102), credit their income tax payments to foreign governments against their U.S. income tax obligations. If they invested in the Western Hemisphere, they also received a reduction of 14 percentage points in their income tax rate.[27]

### NEW ENTRANTS ABROAD

Given these economic and tax incentives, many firms entered (some, like Standard of Indiana, re-entered) the international arena. It has been estimated that 162 U.S. firms moved overseas for the first time between 1945 and 1958.[28] Many were successful in finding oil in a number of countries. One of these areas was the Neutral Zone between Kuwait and Saudi Arabia, where concessions were granted to the American Independent Oil Company and the Getty Oil Company,[29] which entered the Neutral Zone after receiving State Department backing. They were examples of controlled entry.

In other parts of the world, and especially in Latin America, entry was relatively uncontrolled. Smaller American firms found oil in countries as diverse as Venezuela (which in 1956 reversed its 1945 decree prohibiting the granting of new concessions, and opened up its territories to the internationals and newcomers alike[30]), Libya, Nigeria, Australia, Indonesia, Peru, and Ecuador.[31] By the end of the 1950s these newcomers, who had had virtually no foreign holdings at the close of World War II, owned approximately 30 per cent of the overseas reserves of

American companies.[32] In Venezuela alone they had found almost 3 billion barrels of reserves by 1962 and produced some 400 thousand barrels per day during that year.[33] The result of all this activity was the creation of a worldwide "surplus" of oil, a surplus in the sense that far more oil could be produced than could be absorbed by the market at the prices then prevailing.

## The Surplus of Oil

This surplus had the potential of creating both opportunities and problems for the international oil companies and the U.S. government. As long as the surplus remained under their control, they could allocate world oil production as they desired. This allocative power was especially important to U.S. policy planners, who, as was discussed in Chapter 5, wanted to supply Europe with Eastern Hemisphere, and particularly with Middle Eastern, oil and to supply the United States with Western Hemisphere oil. Given a surplus in both hemispheres, such a division of production was possible. U.S. policy attempted to effectuate this division.

### *Surplus as a Means of Control*

The international oil companies, regardless of nationality, welcomed the surplus as a means of enhancing their bargaining power *vis-à-vis* the host countries. As long as there was a surplus, the companies could threaten to reduce production in a recalcitrant country since they would be able to increase production in those countries whose governments were more "reasonable." The surplus was the means through which the international companies were able to skim off a high portion of the economic rents produced by oil.

Iran under Mossadegh was an example of how the surplus was used against a host country. Blair points out that despite the "virtually complete loss of Iran's output" during the years 1951–54, production in the OPEC countries in those years rose at a steady rate of 9.55 per cent per year, *i.e.*, at precisely the same rate as production rose in the entire period from 1950 to 1972.[34] The shortfall in Iranian production was more than offset by substantial increases in production in Kuwait, Saudi Arabia, and Venezuela.[35] If this surplus capacity had not existed, the world would have been forced to come to terms with Mossadegh. Probably more than any other single factor, this surplus was responsible for the overthrow of Mossadegh. At that time OPEC's reserves and production were under the complete control of the international oil companies, and this control enabled them to win in Iran.

The U.S. government, for its part, welcomed a surplus in that, aside from purely strategic reasons, it enabled the government to resist demands by the governments of host countries. A 1945 State Department memo, referring to financial demands by the Saudi government, warned that it "is important that no precedent be created which might lead governments of other countries in which American companies control strategic raw materials to expect that they can count upon U.S. Government assistance in meeting their budgetary difficulties."[36] The greater the surplus the less vulnerable was the United States to these demands.

### *Problems of a Surplus*

If all this surplus had been under the control of the U.S. multinational companies, it would have created no problems. There was a coincidence of interests between these companies and the U.S. government on the use of the surplus. The new entrants, however, had interests of their own which clashed with those of the internationals and did not fit into the plans of the government for the geographic allocation of production. These newcomers, eager to make a quick return on their newly-found reserves, stepped up production significantly, shipping most of this incremental production to the United States, thereby threatening the structure of the domestic industry.

#### GROWTH OF U.S. IMPORTS

In the early postwar years the internationals controlled the bulk of crude imports. As Table 6.1 shows, they accounted for three-quarters of such imports in 1948. With U.S. output in the chief producing states limited by prorationing laws to "market demand," any increase in imports compelled the various state agencies to cut back domestic production in order to maintain the current price. Thus, by directly controlling the flow of imports, the internationals indirectly controlled domestic production. Since they had heavy investments in U.S. production, they had a large financial interest in maintaining the "proper" balance between imports and domestic production.

This system also enabled the internationals to maintain their position in the domestic refining market. As long as they were able to control the bulk of imports, they, in effect, controlled the supply going to independent refiners. Independent refiners are those who do not own crude reserves and are therefore forced to buy crude in the open market. By increasing imports and thereby forcing a cutback in domestic production, the internationals were in a position to compel the independent refiners to buy crude from them.

Because of the industry's price structure, most of the profits in oil are made in the production of crude. The internationals have traditionally pursued a policy of keeping the price of crude high relative to that of refined products. The narrow spread between the price of crude and the price of refined products has brought low profit margins to refining, which has discouraged entry by independent companies into refining. In addition, the relatively high price of crude has increased the value of the depletion allowance available to the sellers of crude. The existing independent refiners, who were being forced in the postwar years to buy an increasing portion of their crude from the internationals, were in reality financing their competitors in refining. This situation was somewhat analogous to the drawbacks Rockefeller received from the railroads hauling his competitors' products.

After the newcomers, including many independent refiners seeking their own sources of crude, entered the overseas field, the internationals lost control over imports. As can be seen in Table 6.1, the internationals' share of imports fell from 74 per cent in 1948 to 68 per cent in 1954 and to 37 per cent in 1959. Furthermore, all the increases in imports between 1948 and 1959 came from the newcomers, whose imports rose by 662 per cent. The imports of the internationals climbed by only 60 per cent during those years.

This precipitous decline in market share threatened to undermine the internationals' ability to control domestic production and, with it, the whole price structure that made crude production much more profitable than refining. It meant that independent refiners were purchasing imported crude from independent producers, thereby ending the system whereby the refiners financed their competitors. In addition, it opened up the possibility that these independent refiners might receive their own, relatively cheap foreign crude, thereby reducing their crude input costs. Since a reduction in crude costs would raise refining profits, the independent refiners would be able to compete more effectively with the internationals. Increased profits, furthermore, would attract new entrants into refining, thus undermining the internationals' dominant position in the field.

The internationals were not the only group disturbed by the flow of imports into the United States. The independent domestic producers,

**Table 6.1: Internationals'[a] Share of U.S. Crude Oil Imports, Selected Years**

| *Year* | *Total Imports* (Thousand Barrels per Day) | *Internationals' Imports* (Thousand Barrels per Day) | *Internationals' Share* (%) |
|---|---|---|---|
| 1948 | 353 | 261 | 74 |
| 1954 | 656 | 444 | 68 |
| 1958 | 953 | 433 | 45 |
| 1959 | 1,119 | 418 | 37 |
| Per Cent Change 1948–59 | +217 | +60 | |

a. The international companies were: Gulf, Shell, Socony, Standard (New Jersey), Standard Oil of California, and Texaco. Shell was not an American company.

Sources: For 1948—U.S. House of Representatives, Select Committee on Small Business, *Hearings. Effects of Foreign Oil Imports on Domestic Producers*, 82nd Congress, 2nd Session (U.S. Government Printing Office, Washington, 1950), Part 2, p. 75. For 1954—U.S. Special (Cabinet) Committee to Investigate Crude Oil Imports, "Petroleum Imports, Report to the President of the United States," July 29, 1957, Tables II and III. For 1958 and 1959—U.S. Oil Import Administration, *Monthly Reports*.

*i.e.*, those companies whose producing properties were entirely located in the States and who owned no refineries, were also adversely affected by them. As the flood of imports rose, the state regulatory agencies drastically cut back their "allowables." The Texas Railroad Commission, for instance, reduced the "allowables" from 366 days in 1948 to 194 days in 1954.[37] This meant that a well that had been allowed to produce at full capacity for every day of the year in 1948 could produce at full capacity for only 194 days in 1954; it had to be shut in for the remainder of the year. Because of this policy of reducing domestic production to make room for imports, imports as a per cent of domestic production rose from 6 per cent in 1948 to 14 per cent in 1958.[38]

This shutting in of U.S. production conformed to the policy proposed by Bullitt of keeping U.S. oil in the ground as a strategic reserve, with current needs being filled by imports. Because of the political difficulties involved in pursuing such a course, this policy had never been adopted. Instead the State Department apparently accepted Collado's plan of allocating Middle East oil to Europe, replacing Western Hemisphere oil there, and supplying Western Hemisphere markets from

Western Hemisphere sources. Such a policy would ease the drain on Western Hemisphere reserves without, at the same time, crippling the domestic industry economically.

The actual pattern of imports did not, however, conform to Collado's plan. Middle East oil began to displace Western Hemisphere oil in the United States as well as in Europe. In 1950, for instance, 59 per cent of U.S. imports of crude and unfinished oils came from Venezuela, the chief source of supply in the Western Hemisphere. Only 23 per cent originated in the Middle East. By 1958, Venezuela's share had fallen to 51 per cent, while imports of Canadian oil, which had been a trickle in 1950, rose to 8 per cent of total imports. However, Canadian imports replaced Venezuelan and not Middle Eastern oil. The Middle East share of total imports rose to 31 per cent.[39]

The adverse effect of these Middle East imports on the Western Hemisphere oil industry concerned the U.S. government. It feared that these Middle East incursions would discourage exploration and development within the Western Hemisphere, a region which, unlike the Middle East, could easily be controlled by the U.S. military. A decline in exploration and development in this area would lead to a fall in the quantity of new reserves found there and might reduce the surplus under U.S. control. The U.S. attitude towards this area was perhaps best exemplified by Dr. Arthur Fleming, director of the Office of Defense Mobilization, who stated in 1955 that "it has always been the policy in Government. . . to consider these countries [Venezuela and Canada] and others in this hemisphere as within the U.S. orbit when dealing with defense questions."[40]

### IMPORT CONTROL PROGRAM

Thus, the government, as well as the internationals, had an interest in controlling the flow of imports. President Eisenhower, accordingly, issued a proclamation in 1959 imposing mandatory controls on oil imports into the United States.[41] This restrictive measure was imposed at the very same time the United States was actively espousing trade "liberalization" throughout the world. It was justified on the grounds of "national security." Imports were presumably affecting national security through their adverse impact on domestic oil exploration and development.

These import controls had important repercussions both domestically and internationally. Domestically, they strengthened the position of the internationals. First of all, they eased the downward pressures on

**Table 6.2: Internationals' Share of U.S. Crude Oil Imports, 1959–67**

| *Year* | *Total Imports* (Thousand Barrels per Day) | *Internationals' Imports* (Thousand Barrels per Day) | *Internationals' Share* (%) |
|---|---|---|---|
| 1959 | 1,119 | 418 | 37 |
| 1960 | 1,151 | 435 | 37 |
| 1961 | 1,201 | 478 | 40 |
| 1962 | 1,303 | 532 | 41 |
| 1963 | 1,325 | 523 | 39 |
| 1964 | 1,404 | 561 | 40 |
| 1965 | 1,472 | 581 | 39 |
| 1966 | 1,485 | 562 | 38 |
| 1967 | 1,398 | 569 | 41 |
| Per Cent Change, 1959–67 | +25 | +36 | |

Source: U.S. Oil Import Administration, News Releases.

domestic crude prices, thereby reducing the opportunities for independent domestic refiners to expand their output and for newcomers to enter refining. The squeeze on the independent refiners was mitigated somewhat by an oil exchange program, which allowed these refiners some access to foreign crude. Nevertheless, this access was not as large as it would have been in the absence of controls. Furthermore, if the price of domestic crude had been allowed to fall, all independent domestic refiners, including those who because of their inland geographical location could not import profitably, would have benefited from this reduction in their input costs.

The controls, second of all, enabled the internationals to increase their share of imports. After declining precipitously between 1948 and 1959 (the first year of mandatory import controls), their share first stabilized and then began to rise until 1967, when the outbreak of the Arab-Israeli war changed the world oil picture dramatically. As can be seen in Table 6.2, the pre-import control pattern was completely reversed.

The improvement in the internationals' position was the direct result of the import quota allocation policy adopted by the U.S. government. Under that policy, licences to import crude were granted only to those companies having refining facilities in the United States. At the onset of the mandatory control program, the six international companies operating in the United States controlled more than 40 per cent of the nation's refining capacity.[42] They were thus major beneficiaries of this policy. This bias towards the internationals was reinforced by allowing companies to receive quotas based on their "historical" imports. Since the internationals were the only significant "historical" importers, they received the main benefits from this provision.

Another aspect of the policy of limiting import licences to companies with refining capacity was that it froze out of the U.S. market independent producers who had acquired foreign oil reserves. Since, by definition, independent producers are non-integrated companies, without refining or marketing facilities, they owned no refining capacity in the United States and they were therefore not entitled to a quota allocation. The lack of an allocation made it extremely difficult for them to sell their oil in the U.S. market. As a result, many independents were forced to sell their overseas reserves to firms with U.S. refining facilities, namely the internationals. The main victim of this provision was Superior Oil, the same company that the U.S. State Department had helped freeze out of the Middle East.

Superior, a non-integrated producer, began production in Venezuela in October 1957, on a concession it had received a year earlier. It was a newcomer in that country and had little chance to cement firm marketing arrangements with independent refiners, its main customers, before mandatory import controls were imposed in 1959. As a result, it had to shut in twenty-eight of its fifty-nine wells. Finally, in 1964, it sold its Venezuelan properties to Texaco, which had both an import quota and refineries overseas, including a 235,000 barrels per day refinery in nearby Trinidad.[43] Thus the quota allocation system helped the internationals obtain a greater share of overseas reserves.

It also helped them *vis-à-vis* those previously independent refiners in the States who had acquired foreign oil reserves. Though these companies did receive a quota, it was insufficient to supply all the oil they needed. In addition, they had a difficult time marketing their overseas production in excess of their quota allocation. The internationals, on the other hand, had the option of sending their surplus production to their overseas refining facilities. The impact of import quotas on them was not nearly as great as on the independent refiners.

Another provision of the program, ostensibly designed to aid the independent refiners, turned out to be of considerable help to the internationals. This was the requirement that an import quota be granted to every refining company in the United States, including those which, because of their geographic location, would never import oil. In order to realize the value of this quota, they would exchange it on a barter basis with coastal refiners, who were in a position to utilize imported crude.

These quotas had a value because the restrictions on imports raised the price of domestic crude above that of foreign crude. If there had been no restrictions, the price of domestic crude could not have exceeded the foreign price plus transportation and handling costs. The mandatory control program enabled the domestic price to pierce the ceiling which would have been imposed by unrestricted imports. The difference between the domestic and foreign price was estimated by President Kennedy's Petroleum Study Committee to have been $1 per barrel[44] and by Professor M. A. Adelman to have been $1.25[45] at a time when the domestic price was about $3 per barrel. By trading its import quota to a coastal refinery, an inland refiner could receive a portion of its domestic crude at the import price.[46]

These exchanges provided an incentive for (1) the establishment of close links between certain coastal and inland refiners, which could be interpreted as anti-competitive behaviour, and (2) mergers between refining companies. By buying out an inland refiner, a company with coastal refineries would also obtain its import quotas. All told, the entire system of assigning quotas led to a rash of mergers, which significantly strengthened the position of the internationals.

Among the mergers which probably were designed to take advantage of the program's quota allocation system were: Gulf Oil's acquisition of the Wilshire Oil Co. of Los Angeles and most of Cities Service's mid-continent refining and marketing facilities; Standard Oil (New Jersey)'s acquisition of Tidewater's refinery in Avon, California; Texaco's acquisition of Paragon Oil and the White Fuel Company; Standard Oil of California's purchase of Standard Oil (Kentucky); and Shell's purchase of El Paso Natural Gas Company's refining and marketing facilities and its acquisition of Buckley Brothers, a small New England oil firm.[47]

The oil import program, in addition to strengthening the position of the international oil companies, also favoured Western Hemisphere oil over that from the Middle East. As Table 6.3 shows, the Western Hemisphere share of imports remained stable, at 70 per cent, during the period of import controls while the Middle East share fell from 20 per cent in 1958, the last year before the imposition of mandatory controls, to 8 per cent in 1967. Though the Middle East share in that year was affected by the Arab-Israeli war and the subsequent blockage of the Suez Canal, it had been declining steadily since 1958. The war merely accelerated this decline. The remainder of the imports came from such diverse areas as Africa, Indonesia, and several Latin American countries. This geographic pattern of imports was achieved through (1) the

**Table 6.3: Geographic Origins of U.S. Imports of Crude Oil and Products, 1958–67 (Per Cent of Total Imports)**

| | *Western Hemisphere* | | | | |
|---|---|---|---|---|---|
| *Year* | *Venezuela and Dutch West Indies*[a] | *Canada* | *Total Western Hemisphere* | *Middle East* | *Other*[b] |
| 1958 | 64.7 | 4.9 | 69.6 | 20.3 | 10.2 |
| 1959 | 65.4 | 5.3 | 70.7 | 19.2 | 10.1 |
| 1960 | 65.6 | 6.4 | 72.0 | 17.6 | 10.4 |
| 1961 | 60.9 | 9.7 | 70.6 | 17.8 | 11.6 |
| 1962 | 60.8 | 11.7 | 72.5 | 14.8 | 12.7 |
| 1963 | 59.2 | 11.8 | 71.0 | 13.6 | 15.4 |
| 1964 | 58.2 | 12.6 | 70.8 | 13.4 | 15.8 |
| 1965 | 57.1 | 12.4 | 69.5 | 13.8 | 16.7 |
| 1966 | 54.9 | 15.7 | 70.6 | 12.8 | 16.6 |
| 1967 | 52.7 | 18.7 | 71.4 | 8.3 | 20.3 |

a. Includes products refined at Aruba and Curacao from Venezuelan crude.
b. Includes Africa, Indonesia, and some Latin American countries.

Sources: U.S. Bureau of Mines and U.S. Oil Import Administration.

imposition of less stringent controls on the imports of residual fuel oil and (2) the exemption of Canadian crude imports from all controls.

Residual fuel oil is one of the chief products of a refinery, the other being gasoline, kerosene, and the middle distillates. Of all these products, the demand for residual is the most elastic, *i.e.*, the most responsive to small changes in price. The relatively high elasticity of demand for residual is due to the availability of close substitutes, namely natural gas and steam coal. Residual, unlike the other products, is used mainly as an under-boiler fuel for heating liquids in thermal electric power plants and in large industrial establishments. Since this heating function can be performed by natural gas and coal as well, many utilities and industrial plants have boilers which can switch to these alternative fuels. A slight change in the relative price of these fuels can thus induce a large shift in usage. This price sensitivity has meant that the price of residual fuel oil, on a calorific basis, cannot be higher than the prices of natural gas and steam coal in the relevant markets. Because of U.S. government regulation of gas prices and a declining demand for coal, the prices of these products were low. Refiners wishing to market residual had to price within the ceiling imposed by these fuels.

No such restrictions applied to the other refinery products, whose demand elasticity, at least in the short run, was much lower than that of residual fuel oil. As a consequence, refiners charged a higher price for these products than for residual, and this gave them an incentive to reduce the residual fuel oil yield from a barrel of crude and to increase the yield of the higher-valued lighter products, especially gasoline. Technological innovations, like the cracking process, made such a change in yields possible. As a result, refiners reduced their output of residual fuel oil. Along the East Coast of the United States, where the demand for residual fuel oil was strongest, the output of residual fell from 242,000 barrels per day in 1951 to 191,000 barrels in 1958. This decline was more than offset by an increase in residual imports from 326,000 to 489,000 barrels per day during the same period.[48]

The residual came mainly from refineries in Venezuela and in the Dutch West Indies islands of Aruba and Curacao, which used the heavy crude oil of Lake Maracaibo as an input. Because this heavy crude yields a high percentage of residual fuel oil, its price is lower than the lighter crudes, which yield a relatively high percentage of gasoline. The oil companies hence found it more profitable to produce their residual from the low-priced Venezuelan heavy crudes than from the higher-priced American and Middle East lighter crudes.

In principle they could have refined the heavy Venezuelan crude in the United States, especially since it is normally cheaper to ship crude long distances than to ship products. Venezuelan policy, however, forced them to locate refineries in Venezuela. Because of this policy, refining capacity in Venezuela expanded from 285,000 barrels per day in 1951 to 820,000 in 1959.[49] Faced with this forced expansion of capacity, the oil companies quite logically decided to use it to convert their heavy crude into residual fuel for sale in the United States, while at the same time they reduced the unprofitable residual output of their U.S. refineries. Thus, in the period between 1951 and 1958, the internationals who had refineries both on the East Coast of the United States and in the Venezuelan-Dutch West Indies complex[50] increased their imports of residual fuel by 50 per cent. However, the imports of crude oil from the Venezuelan-Dutch West Indies region rose at a faster rate, by 95 per cent.

After the imposition of mandatory import controls in 1959, the quantity of crude oil imported from Venezuela actually fell. By 1967 it was 2 per cent below the 1958 level. Residual imports from the area, on the other hand, skyrocketed by 115 per cent. The less stringent restrictions on imports of residual induced its substitution for crude, and this led to some important results.

First, it contributed to the continued dependence of the Atlantic Coast region in general, and of New England in particular, on residual fuel oil. If the restrictions on residual imports had been as severe as those on crude, industries in these regions would have switched to natural gas and coal. Second, it provided a partial offset to Venezuela for the loss in income flowing from import controls. The Venezuelan government, while recognizing this benefit, would have preferred that the United States increase its imports of the higher-valued lighter crudes rather than those of the lower-valued heavy crudes. Third, it benefited the international oil companies with refining facilities in the Venezuelan-Dutch West Indies region by allowing them to recover their investments on the refineries built in the mid-1950s. It also enabled the internationals to increase their control over imports, since most newcomers to Venezuela did not have refining facilities in the area. The substitution of product for crude imports thus gave the internationals an important advantage. Fourth, it conformed to the general U.S. policy of directing Western Hemisphere oil to the United States and Middle Eastern oil to Europe. As can be seen in Table 6.3, Venezuela and the Dutch West Indies' share of total imports into the United States fell between 1958 and 1967. This fall would have been considerably steeper if the import program had not encouraged the substitution of residual for crude. Finally, it had an adverse impact on the domestic coal industry. Hard hit by the loss of its traditional markets in home heating and railroads, that industry sought to improve its position through expanding in the fast-growing electricity generation market. The growing imports of residual fuel foreclosed that market along the East Coast, contributing to a drop in coal production and a closing of mines. A similar pattern, with dire consequences for the future, occurred in Europe.

Despite the drop in the Venezuela and Dutch West Indies share, the Western Hemisphere share remained stable because the rise in the Canadian share offset the fall in the Venezuelan. Canada's share rose because the U.S. government exempted imports from Canada from the control program; an amendment to the original import proclamation removed all barriers on imports "which are transported into the United States by pipeline, rail, or other means of overland transportation from the country where they were produced."[51]

It is significant that the original proclamation did not provide for such an exemption. It would have seemed logical to grant one in a program ostensibly based on protecting national security. Canadian oil could, after all, provide a safe alternative fuel source in the event of a national emergency. The importation of Canadian oil furthermore conformed to the objective of favouring Western Hemisphere oil over that of the Middle East. Despite the obvious advantages of Canadian oil, the U.S. was reluctant to grant Canada any exemption. The exemption was granted in the end only because the Canadian government insisted on it and had enough bargaining power to force the U.S. to accede to it.

However, before discussing the specific events leading to the exemption of Canada, we should first examine the way petroleum had developed in Canada and the implications of this development for U.S. policy. This is done in the next chapter.

### Notes

1. See for instance Galbraith's theory of "Countervailing Power" in his *American Capitalism*, Sentry ed. (Houghton Mifflin, Boston, 1962), pp. 108–34.
2. W. J. Kennedy (ed.), *Secret History of the Oil Companies in the Middle East* (2 vols., Documentary Publications, Salisbury, N.C., 1979), vol. 1, p. 111.
3. In his report to the State Department, Childs stated that he "was not impressed at all by the group from the Superior Oil Company nor was, I may add, the crew of the private plane transporting them. Members of the crew, who have some familiarity with this area, described the party to a member of the Legation staff as a group of joy-riding promoters travelling on a junket at the expense of the company." See ibid.
4. Ibid., pp. 110–11.
5. Ibid., pp. 107–8.
6. Ibid., p. 162.
7. Ibid., p. 96.
8. Ibid., p. 199.
9. Ibid., p. 161.
10. Ibid., p. 204.
11. Ibid.
12. Ibid., p. 155.
13. Ibid., p. 217.
14. Ibid., p. 219.
15. Ibid., p. 220
16. Ibid., vol. 2, p. 382.
17. C. Issawi & M. Yeganeh, *The Economics of Middle Eastern Oil* (Praeger, New York, 1962), pp. 178–79.
18. Kennedy, *Secret History of the Oil Companies in the Middle East*, vol. 1, p. 139.
19. Ibid., p. 138.

20. *Petroleum Facts and Figures: Centennial Edition* (American Petroleum Institute, New York, 1959), p. 240.
21. Ibid., pp. 246–47.
22. Ibid., p. 385.
23. U.S. Department of Commerce, *Business Statistics: 1963 Edition* (U.S. Government Printing Office, Washington, 1963), p. 101.
24. *Petroleum Facts and Figures: Centennial Edition*, p. 22.
25. *Petroleum Imports, A Report to the Secretary of the Interior* (National Petroleum Council, Washington, 1955), p. 51.
26. *Petroleum Facts and Figures: Centennial Edition*, p. 213.
27. U.S. Senate, Subcommittee on Multinational Corporations of the Committee on Foreign Relations, *Hearings, Multinational Corporations and United States Foreign Policy*, 93rd Congress, 2nd Session (U.S. Government Printing Office, Washington, 1974), part 4, p. 117.
28. L. Fanning, *The Shift of World Petroleum Power Away from the United States* (Gulf Oil Company, Pittsburgh, 1958), p. 33.
29. Issawi & Yeganeh, *The Economics of Middle Eastern Oil*, pp. 178–79.
30. "A Review of Venezuelan Oil Activities during 1956," *Venezuela Up-to-Date*, vol. 7 (July–August 1957), p. 14.
31. N. Jacoby, *Multinational Oil, A Study in Industrial Dynamics* (Macmillan, New York, 1974), pp. 129, 134–35, 138.
32. A. Gols, "United States Foreign Oil Investments," unpublished Ph.D. thesis, University of Oregon, 1961, p. 21.
33. Jacoby, *Multinational Oil*, p. 138.
34. J. Blair, *The Control of Oil* (Pantheon, New York, 1976), p. 101.
35. Ibid., p. 100.
36. Kennedy, *Secret History of the Oil Companies in the Middle East*, pp. 105–6.
37. The Railroad Commission of Texas, *Annual Report(s) of the Oil and Gas Division*.
38. E. Shaffer, *The Oil Import Program of the United States: An Evaluation* (Praeger, New York, 1968), p. 19.
39. Ibid., p. 128.
40. *The Oil and Gas Journal*, Nov. 7, 1955, p. 71.
41. "Adjusting Imports of Petroleum and Petroleum Products into the United States," U.S. Presidential Proclamation 3279, March 10, 1959 (24 F.R. 1781).
42. U.S. House of Representatives, Select Committee on Small Business, *Hearings, Small Business Problems Created by Petroleum Imports*, 87th Congress, 1st Session (U.S. Government Printing Office, Washington, 1962), pp. 975–91.
43. Shaffer, *The Oil Import Program of the United States*, pp. 152–53.
44. U.S. Petroleum Study Committee, "A Report to the President," Sept. 4, 1962, p. 2.
45. M. A. Adelman, "Efficiency of Resource Use in Crude Petroleum," *The Southern Economic Journal*, vol. 31 (1964), p. 103.
46. For a detailed explanation of how the exchange system operated, see R. Manes, "Import Quotas, Prices and Profits in the Oil Industry," *The Southern Economic Journal*, vol. 30 (1963), pp. 13–24.

47. Shaffer, *The Oil Import Program of the United States*, pp. 180–82, 205, 208.
48. Ibid., p. 89.
49. *Petroleum Facts and Figures: Centennial Edition*, pp. 438–39.
50. Four of the internationals—Standard Oil (New Jersey), Gulf, Standard Oil of California, and Texaco—controlled approximately 45 per cent of refining capacity on the East Coast and 65 per cent in the Venezuelan-Dutch West Indies area. See Shaffer, *The Oil Import Program of the United States*, p. 91.
51. U.S. President, "Modifying Proclamation No. 3279 of March 10, 1959, Adjusting Imports of Petroleum and Petroleum Products," Proclamation 3290, April 20, 1959 (24 F.R. 3527).

# CHAPTER SEVEN

# The Control of World Oil: The Role of Canada

At the end of World War II Canada was producing only small quantities of oil. Production from its main source, the Turner Valley, was rapidly declining. In 1946 more than 90 per cent of domestic demand for petroleum had to be met by imports. Even the Prairie refineries, the ones closest to the Turner Valley, had to import some of their crude from Texas and Oklahoma at transportation costs at least four times higher than those paid by refineries in the United States.[1]

Canada's oil picture changed dramatically in 1947 after the discovery by Imperial Oil of a large field at Leduc, Alberta, a few miles south of Edmonton.[2] This discovery led both to a rush to find new oil fields and to a significant increase in domestic production. Following Leduc, a number of important new discoveries were made. Among them were Redwater in 1948,[3] Pembina in 1953,[4] and Rainbow Valley in 1965.[5] With these discoveries Canada entered the world of oil. The discoveries gave Canada the possibility of becoming self-sufficient in oil and of turning into a net exporter of petroleum.

It the years following Leduc, Alberta's production increased by, literally, leaps and bounds. As can be seen in Table 7.1, production rose from approximately 7.7 million barrels in 1946, the last full year before the Leduc discovery, to 10.5 million in 1948, the first full year after Leduc. Production continued to make spectacular gains in subsequent years until it reached a temporary peak of 144 million barrels in 1956. All told, between 1946 and 1956, production rose by almost 1800 per cent or at an average annual rate of 34 per cent. After 1956 Alberta encountered serious problems marketing its crude. These problems will be discussed later.

## Continental Energy

Until these problems arose, Alberta found increased markets for its oil by sending it to the other Prairie provinces, British Columbia, Ontario, and the United States. These markets were served by a pipeline network, which was—and still is—controlled by the large, multinational oil companies.

Shortly after the Leduc discovery, Imperial became convinced that it had more oil than it could dispose of in Alberta. It therefore decided to build a 720-kilometre pipeline from Edmonton to Regina to supply its refinery in the latter city.[6] However, before Imperial could build this line, it had to overcome one important political problem, a problem inherent in the federal structure of the Canadian state. Because the line was to traverse provincial boundaries, it could be subject to different, and often conflicting, sets of provincial regulations. There was, furthermore, always the possibility that a province could prohibit the construction of a pipeline within its territory.

Imperial thus sought to have interprovincial pipelines placed under federal jurisdiction. In 1949, it applied to the federal government for the incorporation of the proposed line. Parliament then passed the Pipelines Act of Canada, which gave Ottawa exclusive jurisdiction over all interprovincial pipelines in the country.[7]

**Table 7.1: Alberta Crude Oil Production, 1946–56**

| *Year* | *Production* (Millions of Barrels) | *Per Cent Change over Previous Year* |
|---|---|---|
| 1946 | 7.7 | — |
| 1948 | 10.5 | 36.4 |
| 1949 | 19.8 | 88.6 |
| 1950 | 27.1 | 37.6 |
| 1951 | 45.8 | 69.0 |
| 1952 | 58.8 | 28.4 |
| 1953 | 76.7 | 30.4 |
| 1954 | 87.6 | 14.2 |
| 1955 | 112.9 | 28.9 |
| 1956 | 143.7 | 27.3 |
| Change, 1946–56 | 136.0 | 1,766.2 |
| Per Cent Change per Year, 1946–56 | | 34.0 |

Sources: For 1946—calculated from E. Gray, *The Impact of Oil* (The Ryerson Press/Maclean-Hunter Ltd., Toronto, 1969), p. 14; all other years—*1971 Statistical Yearbook—Canadian Petroleum Association* (Canadian Petroleum Association, Calgary, 1972), p. 75.

*Interprovincial Pipe Line*

Given this federal protection, Imperial proceeded with plans to construct the pipeline, fittingly called the Interprovincial Pipe Line (IPL). Before construction started, Imperial decided it would use the line as a means of providing its refinery at Sarnia (which was then supplied with prorationed oil from the U.S. mid-continent fields[8]) with Alberta oil, which at that time was not prorationed.[9] The pipeline was to be extended from Regina to Gretna, on the Manitoba–U.S. border, and from there to Superior, Wisconsin. Tankers would then take the oil from Superior over the Great Lakes to Sarnia. To facilitate movement, IPL organized a wholly-owned U.S. subsidiary, the Lakehead Pipeline Company, which would handle operations in the U.S.[10]

In September 1949 the Board of Transport Commissioners gave IPL permission to build the extension to Superior.[11] This decision sparked a heated debate in Parliament. The Progressive Conservative opposition demanded an all-Canadian route, terminating in the Thunder Bay area. C. D. Howe, the minister of trade and commerce, defended the decision, claiming that the Superior route was more economical[12] and that it would provide access to U.S. markets in the Chicago area. Howe refused to override the decision[13] and a year later, in 1950, oil was being transported from Edmonton to Sarnia.[14] In 1953 the line was extended across Wisconsin and Michigan to Sarnia, eliminating the need to use the Great Lakes. In 1956 it was extended to Port Credit, a suburb of Toronto. With this extension, the line stretched over three thousand kilometres, which made it the longest crude oil pipeline in the non-Communist world.[15] In 1967, the line was looped around the Chicago area, opening that region to Canadian oil.[16]

The routing of this line, which provided Toronto with Alberta oil, had two important implications. First, it subjected domestic Canadian oil to U.S. control. The U.S., if it ever so desires, has the power to cut the flow of Canadian oil from Alberta to Toronto. This power can give the U.S. strong leverage in bargaining with Canada and is a means by which the U.S. can enforce its discipline on Canada. The political implications of this route raise questions as to whether it was chosen by Imperial solely for economic reasons. Given the close relationship between Imperial's parent, Standard Oil (New Jersey), and the U.S. State Department, one wonders whether these political considerations were not also an important factor in Imperial's insistence on this particular route.

The second implication was that this route also gave U.S. refineries direct access to Canadian crude reserves. It thus opened up the possibility of Canadian oil being integrated with U.S. oil in a continental energy policy.

*Trans Mountain Oil Pipeline*

The opportunities for this integration were enhanced by the construction of the Trans Mountain Oil Pipeline, which connects Edmonton with Vancouver and the Puget Sound area in the state of Washington.[17] This line, like the Canol line in the Northwest Territories, was built to serve the needs of the U.S. armed forces. Prior to the opening of the Trans Mountain line, the Puget Sound area had no refineries. All its oil requirements were provided by California refineries. In 1951, during the Korean War, the U.S. Petroleum Administration for Defense grew concerned that the expected growth in California reserves would be insufficient to serve the Pacific Northwest area. It looked upon Alberta oil as an alternative. Rejecting other possible sources, such as the Middle East, the Petroleum Administration for Defense emphasized "the desirability of making available to a [sic] market as much producible crude oil in the Western Hemisphere as possible."[18] This reasoning was in line with Collado's plan of keeping Western Hemisphere oil within that hemisphere and Eastern Hemisphere oil in the rest of the world, an allocation of world oil supplies which, at that time, best served the interests of the United States.

The U.S. explicitly stated that there were "no practical means" of realizing its "desired defense considerations other than an Alberta–Puget Sound pipeline."[19] The Petroleum Administration for Defense recommended that the line be built along with "the construction of increased refining capacity."[20] It also promised to assist in the "acquisition of steel to this end."[21] Though steel was scarce at that time, it helped obtain steel and other scarce materials for both the line itself and the oil industry in western Canada.[22]

Oil-producing firms in Alberta, both domestic and foreign-owned, were examining the possibilities of building a line to serve West Coast markets for purely commercial reasons, but they probably would not have built it at that particular time. As the second report of the Borden Commission (Royal Commission on Energy) notes, it was the "outbreak of the Korean War and defence considerations" that "hastened the decision to build the pipe line."[23] With the construction of the pipeline, Canadian oil thus became part of the U.S.'s strategic oil reserves. Another link was forged in the creation of a continental energy policy.

In this connection it is interesting to note that five firms applied to the Board of Transport Commissioners in Ottawa for permission to build the line. For various reasons, four of the firms withdrew their applications.[24] Whether these withdrawals were voluntary or the result of "persuasion" is a matter of conjecture. The only remaining applicant, Trans Mountain Oil Pipe Limited, received permission to build and operate the line. The two major owners of Trans Mountain at the time of its incorporation in 1951 were Imperial Oil and Canadian Bechtel Limited,[25] which actually built the line.

The parents of both these firms, Standard Oil (New Jersey) and the Bechtel Corporation, respectively, have had close ties with the U.S. government for years. The ties of Standard Oil have already been discussed and will be discussed in more detail in subsequent chapters. Bechtel's continuing ties with the U.S. government[26] will also be detailed later (see Chapter 11).

Presumably because it was more interested in building the line rather than operating it, Bechtel did not remain a major shareholder for very long. Shortly after its incorporation Trans Mountain issued 1.5 million shares of stock. Of this amount, almost half—670,000—were purchased by six oil companies with refinery facilities on the West Coast, namely, Imperial Oil, Shell, Standard Oil of British Columbia (a subsidiary of Standard Oil of California), Gulf Oil, Union Oil, and Richfield Oil. Other oil companies bought another 250,000 shares. All told, oil

**Table 7.2: Canadian Crude Production and Exports, 1951–60**

| *Year* | *Production* (Millions of Barrels) | *Exports* (Millions of Barrels) | *Exports as a Percentage of Production* (%) |
|---|---|---|---|
| 1951 | 47.5 | 0.3 | 0.6 |
| 1952 | 61.2 | 1.4 | 2.3 |
| 1953 | 80.8 | 2.5 | 3.1 |
| 1954 | 96.0 | 2.3 | 2.4 |
| 1955 | 129.3 | 16.7 | 12.9 |
| 1956 | 171.8 | 42.4 | 24.7 |
| 1957 | 181.1 | 55.3 | 30.5 |
| 1958 | 164.7 | 29.7 | 18.0 |
| 1959 | 183.6 | 33.7 | 18.4 |
| 1960 | 189.5 | 41.3 | 21.9 |

Source: *1971 Statistical Yearbook—Canadian Petroleum Association* (Canadian Petroleum Association, Calgary, 1972), pp. 75, 101.

**Table 7.3: Destination of Canadian Crude Exports to the United States, 1955–58**

| *Year* | *West Coast* (Millions of Barrels) | (%) | *Mid-West* (Millions of Barrels) | (%) | *Total* (Millions of Barrels) |
|---|---|---|---|---|---|
| 1955 | 11.3 | 66 | 5.8 | 34 | 17.1 |
| 1956 | 25.4 | 58 | 18.3 | 42 | 43.7 |
| 1957 | 34.2 | 62 | 21.0 | 38 | 55.2 |
| 1958 | 8.9 | 28 | 22.4 | 72 | 31.3 |

Calculated from: Canada, Royal Commission on Energy, *Second Report* (Ottawa, 1959), pp. 32 and 37.

companies owned 920,000 shares. The four largest shareholders, each with 130,000 shares, were Imperial, Shell, Standard Oil of British Columbia, and Gulf.[27] The control of the pipeline thus remained firmly in the hands of these multinational oil companies, all but one of which, Shell, were American.

The building of the pipeline induced the construction of refineries in the Puget Sound area. In 1954 the General Petroleum Corporation, a subsidiary of Mobil, completed a 35,000 barrels per day refinery, and in 1955 Shell opened a 55,000 barrels per day refinery in the region.[28] Following the construction of these refineries, Canadian exports started rising sharply. As can be seen in Table 7.2, exports, which had previously consisted of marginal quantities going to refineries located on the U.S. portion of the IPL, jumped sharply in 1955. They reached a peak in 1957, only to fall abruptly in the following year. At their peak exports accounted for almost one-third of Canadian production.

Most of the exports were carried over the Trans Mountain line to West Coast refineries. As Table 7.3 shows, refineries in the Pacific area absorbed approximately three-fifths of crude exports in the years 1955–57. During the years 1956 and 1957 the exports went beyond the Puget Sound region to California, a movement triggered by the escalation of tanker rates following the closure of the Suez Canal during the 1956–57 Arab-Israeli war. The high tanker rates raised the delivered cost of Middle East and Indonesian crudes in California above that of Alberta crude.

### MARKET CHANGES

With the reopening of the Suez Canal in the latter part of 1957, tanker prices plummeted and Alberta oil lost its advantage in California. Given the temporary nature of the Suez crisis, this loss was not surprising.

What was unanticipated was the simultaneous loss of the Puget Sound market, which was designed to run on Alberta oil. The loss in that market was almost as dramatic as the one in California. In July 1957 the Puget Sound refineries consumed 94,000 barrels per day of Alberta crude. By December 1958 consumption had fallen to 10,000 barrels per day.[29]

Alberta would have lost even this 10,000 barrels a day market had not Mobil, the company which purchased the oil, made a special deal with Imperial under which Mobil would purchase Imperial's Alberta oil for use in its Puget Sound refinery and, in return, Imperial would purchase Mobil's Venezuelan oil for use in its Montreal refinery.[30] If it had not been for this arrangement, there is little doubt that Mobil would have joined Shell, the other refiner in the Puget Sound area, in dropping entirely its purchases of Canadian crude.

There were a number of reasons for this change in attitude towards Alberta oil. Both the Trans Mountain line and the refineries, as has already been mentioned, were planned during the Korean War. The military needs of the United States were paramount in the initiation of these projects. Linked with these needs was the apprehension that the growth in California reserves would be insufficient to satisfy both the military and civilian requirements of the Pacific North-West.

By the time the Suez Canal reopened, the world oil picture had changed significantly. The Korean conflict had ended several years earlier and the United States was not engaged in any other major conflict. The military requirements for oil were correspondingly reduced. In addition the United States was in the throes of a recession, which lowered the civilian demand for oil. Third, reserves in the United States and in the rest of the world grew much faster than was earlier anticipated. An oil surplus thus reappeared in the U.S. and in the rest of the world.

Shell, for instance, had a large inventory of crude in California. It also had long-established agreements, which it did not want to break, with California producers. Furthermore, its refineries in California, which were considerably older than the one in Puget Sound, were less efficient. Shell therefore reduced its refinery output in California, sent its California crude for processing in the state of Washington, and completely eliminated its purchases of Alberta crude.[31]

Texaco, in effect, did the same thing. In 1958, when it opened its Puget Sound refinery, which was originally designed to be supplied by Trans Mountain, it fed it with domestic rather than Canadian crude.[32]

Aside from the purely economic considerations, a factor that influenced Texaco and the other two refiners was the imposition by the U.S. government of "voluntary" quotas on oil imports, quotas imposed on Canadian as well as on all other non-U.S. oil.[33] U.S. policy thus provided these refiners with strong incentives against the use of Canadian oil.

The impact of these restrictions was especially severe on Alberta. The steady rise in Alberta's production turned, after 1956, into a steep decline. As can be seen in Table 7.4, Alberta's production dropped by 5 per cent in 1957 while national production rose by 5 per cent. In 1958 production in Alberta fell by almost 20 per cent, in Canada as a whole by less than 10 per cent. Between 1956 and 1958, Alberta's share of national production declined from 84 to 68 per cent.

The Alberta downturn highlighted the risks inherent in an export-oriented resource policy. While there is normally a risk in doing any business at all—a domestically-oriented resource policy is also not risk free—the risks in exports are larger, especially if the bulk of the exports go to one country and if, also, the private importers are large, multinational corporations who have access to a number of alternate supply sources.

In such an oligopsonist setting the few large buyers can play one group of sellers off against the other. This is precisely what Rockefeller had done to the small producers in Pennsylvania. And this, in effect, was what the large international oil companies did to the producers in Alberta almost a century later.

**Table 7.4: Crude Oil Production in Alberta and Canada, 1956–61**

| | ALBERTA | | CANADA | | *Alberta as Percentage of Total Canadian Production* |
|---|---|---|---|---|---|
| *Year* | *Production* (Millions of Barrels) | *Change over year* (%) | *Production* (Millions of Barrels) | *Change over year* (%) | (%) |
| 1956 | 143.7 | — | 171.8 | — | 83.6 |
| 1957 | 136.8 | - 4.8 | 181.1 | 5.4 | 75.5 |
| 1958 | 112.5 | -17.8 | 164.7 | -9.1 | 68.3 |
| 1959 | 128.8 | 14.5 | 183.6 | 11.5 | 70.2 |
| 1960 | 130.5 | 1.3 | 189.5 | 3.2 | 68.9 |
| 1961 | 157.8 | 20.9 | 220.8 | 16.5 | 71.5 |

Source: *1971 Statistical Yearbook—Canadian Petroleum Association* (Canadian Petroleum Association, Calgary, 1972), pp. 75.

On a much larger scale the U.S. government was able to use this dependence for its own ends. When it felt that it needed Canada's oil for its military adventures, it left no stone unturned to obtain Canada's oil. When that need no longer existed, the U.S. had no hesitation in leaving Canada, which had grown economically dependent on oil exports, high and dry. In 1955, when the U.S. started its first "voluntary" import control program, the government of Canada attempted to remind the U.S. that "the United States on strategic grounds gave support to the construction of the Trans Mountain pipeline and for the construction of additional refining capacity in the State of Washington,"[34] but this fell on deaf ears. In the late 1950s the producers in Alberta, and the people of Canada, learned all too painfully the dangers of becoming economically dependent on a nation whose bargaining power far exceeded theirs.

IMPORTS

Faced with declining markets for Alberta oil, the independent producers in that province demanded that the federal government protect domestic producers by restricting the imports of foreign oil. However, this would have involved changing the market division established by the internationals. Because of the pricing system adopted by Imperial and followed by all the other major companies, the price of Canadian crude was set so that it would equal the price of Illinois Basin crude in the Ontario market.[35] This price precluded Alberta oil from Montreal, Canada's largest refining centre,[36] which was supplied exclusively by foreign oil, mainly from Venezuela.[37] Unlike Alberta, Montreal was not adversely affected by the events following the reopening of the Suez Canal. Thus, as Table 7.5 shows, while Alberta production fell drastically (approxi-

**Table 7.5: Crude Oil Production in Alberta and Imports into Canada, 1956–61 (Millions of Barrels)**

| | *Alberta* | | *Imports* | |
|---|---|---|---|---|
| *Year* | *Production* | *Change* | *Production* | *Change* |
| 1956 | 143.7 | — | 106.3 | — |
| 1957 | 136.8 | -6.9 | 111.9 | 5.6 |
| 1958 | 112.5 | -24.3 | 107.4 | -4.5 |
| 1959 | 128.8 | 16.3 | 116.3 | 8.9 |
| 1960 | 130.5 | 1.7 | 126.8 | 10.5 |
| 1961 | 157.8 | 27.3 | 133.2 | 6.4 |

Source: *1971 Statistical Yearbook—Canadian Petroleum Association* (Canadian Petroleum Association, Calgary, 1972), pp. 75, 105.

**Table 7.6: Refining Companies and Capacities, Montreal 1957**

| *Company* | *Capacity* (Thousands of Barrels per Day) | *Per Cent* |
|---|---|---|
| Imperial | 71.8 | 28 |
| Shell | 60.0 | 23 |
| Texaco | 59.0 | 23 |
| Gulf | 45.0 | 18 |
| Petrofina | 20.0 | 8 |
| Total | 255.8 | 100 |

Source: Canada, Royal Commission on Energy, *Second Report* (Ottawa, 1959), p. 95.

mately 31 million barrels between 1956 and 1958), imports, mainly to Montreal, increased by approximately 1 million barrels.

At that time Montreal had five refineries, all owned by the international oil companies. (The refineries and their capacities are listed in Table 7.6.) In 1957 these five firms controlled 80 per cent of Canadian refining capacity and produced about 43 per cent of the domestic crude.[38] Four of the companies—Standard Oil (New Jersey), Gulf, Royal Dutch-Shell, and Texaco—had more than an 85 per cent interest in total Venezuelan crude production.[39] These companies were thus in a position to play off Venezuela against Alberta, if they so desired.

That they preferred Venezuelan and other oil to Albertan was quite evident. As was already noted, an affiliate of Shell completely stopped its purchases of oil flowing over the Trans Mountain line. The Puget Sound refinery of Texaco, upon its opening in 1958, did not take any Albertan crude. Texaco, in addition, had a refinery which was in the marketing area of the U.S. section of the IPL; however, that refinery also did not buy any Canadian crude. This was true, as well, of Gulf's refinery in that same area.[40] These latter two refineries opted for domestic U.S. crude.

*Alberta-Montreal Pipeline*

Faced with this refusal by the internationals to use Alberta oil either in the United States or in Montreal, the independents demanded that a pipeline be built to connect Montreal with Alberta's oil fields. In actuality there were three distinct proposals. One came from Home Oil, at that time Canada's largest independent oil company, which called for the construction of a pipeline paralleling the route of the IPL, from Edmonton to Montreal.[41] Home had earlier organized a pipeline company, the Independent Oil Pipe Line Company, to carry crude to the West Coast,[42] and this company was one of the applicants which had withdrawn their bids in favour of Trans Mountain Pipe Line Company. The construction of an additional line to the East would have broken Imperial's monopoly, through its control of IPL, of traffic to the East. It would have also given Home considerable control over production in western Canada.

For this reason, while some other independents had reservations about supporting the Independent Pipe Line Company, preferring instead to back a "citizens' line" with a much wider distribution of ownership,[43] they nevertheless wound up defending the line in hearings before the Borden Commission,[44] a royal commission established by Prime Minister John Diefenbaker to examine the problems facing domestic oil producers.[45]

The second proposal submitted to the commission was that of IPL. Not surprisingly it opposed the construction of a competing parallel line, proposing instead the extension of the existing IPL line from Sarnia to Montreal.[46]

A third proposal was the construction of a government-owned line along an all-Canadian route. In the words of the Borden Commission "such a pipeline might be regarded as a transportation facility created to serve the national interest."[47]

Before making its decisions on these three proposals, the Borden Commission appointed Canadian Bechtel, the same company which, along with Imperial, had been the original partner in the Trans Mountain pipeline, to estimate the costs of each project. Bechtel found the extension of the IPL the cheapest alternative and the all-Canadian line the most expensive.[48] In contrast, the pipeline construction firm of Dutton-Williams, hired by Home Oil and using different assumptions from Bechtel, found the Independent Oil Pipe Line would be the least expensive.[49]

Recognising that different assumptions about the extent of debt financing, interest rates, depreciation schedules, and income tax payments "play an important role in the determination of the cost of transmitting oil by pipeline,"[50] the commission asked Bechtel to estimate the cost of a government pipeline which would traverse an all-Canadian route and which would be entirely financed by 5 per cent bonds, would depreciate at 2 per cent per year, would operate as a non-profit corporation, and would pay no income taxes. Bechtel reluctantly concluded that such a line would be, by far, the cheapest alternative. The author of the report inserted an apologetic addendum, which read in part:

> . . . the basis of calculation you have asked me to use is not similar to a normal commercial approach, the conclusions reached are not directly comparable with estimates prepared for a normal type of development. I know you fully appreciate the differences, but when an engineer follows an unusual financial approach in estimating the cost of some process or act, you will understand how anxious he is lest the figure so prepared be used in some way in direct comparison with the estimated cost of the same process when using a normal financial approach.[51]

The commission accepted these caveats, adding that "the question of public ownership of such a pipe line involves considerations beyond the question of securing lower transportation costs to permit the Montreal market to be reached by Canadian crude."[52] Though the commission did not specify what these other "considerations" were, one can assume they were of an ideological nature and that they may also have involved relations with the international oil companies and the U.S. government.

These considerations had surfaced in the controversy several years earlier over the financing of the Trans-Canada natural gas pipeline. This controversy toppled the government of Louis St. Laurent and brought about the election of John Diefenbaker. Because the St. Laurent government stipulated that this line traverse an all-Canadian route, its backers asked the government for a subsidy to compensate them for the difference in costs between that route and a cheaper one going through the United States. The government agreed to do this in return for common stock and convertible debentures, which, upon conversion, would give the government full control over the line.

The U.S.-controlled firms in Alberta objected to selling their gas to a line in which the government had a controlling interest, despite the urgings of Alberta's Social Credit government for them to do so.[53] The one major exception was Gulf Canada, which was to supply Trans-Canada with about 40 per cent of its gas. Gulf Canada agreed to supply this quantity provided that it received permission from its parent in Pittsburgh. The parent categorically refused to do so. This refusal, as Crane has pointed out, had profound implications for Canada's energy policy.

> What this meant was that a project negotiated with the government of Canada and approved by the governments of Alberta and Ontario was being vetoed by a U.S. oil company. Without the Gulf gas contract, Trans-Canada could not arrange the necessary private-sector financing and delivery contracts to place its orders for steel pipe.[54]

The Borden Commission seemingly did not want to challenge the international oil companies on this score. Neither, as Crane notes, did the governments of Canada and Alberta.[55] The federal government wound up by giving financial assistance to the natural gas pipeline, the control of which had passed into the hands of three U.S. companies—Tennessee Gas Transmission, Gulf Oil, and Continental Oil.[56] The furore raised by this assistance brought down the St. Laurent government.

Aside from the question of a publicly-owned pipeline, the Borden Commission report raised some important issues. First, it pointed out that there are no clear, unambiguous estimates of the costs of a large-scale project: the cost estimates are a function of the assumptions made by the consultants, and the assumptions are usually a function of the interests of the consultants' client.

Second, the qualms of the Bechtel consultant about not "using a normal financial approach" pose the question of what is "a normal financial approach." Such an approach presumably guarantees the private investor a specified rate of return. That that rate of return may be quite high can be inferred from the differences between the estimates for transmission costs of the extension of the IPL, and those of the government-owned line. Bechtel placed the former at 69.1 cents per barrel[57] and the latter at 49 cents per barrel.[58] In other words, the estimated costs of the cheapest private line were over 40 per cent higher than those of a government-owned one.

That the profits on pipeline operations may have been quite substantial has been indicated by a report issued almost twenty-five years later by the director of Investigation and Research, Combines Investigation Act (Bertrand Report). In preparing this report the director seized many internal documents of the oil companies, including documents on the profitability of pipeline operations. One of these documents was the 1957 report by E. F. Turner of Imperial Oil entitled "Canadian Pipe Lines—Corporate Finance and Earnings." Turner commented that "there were instances of rapid depreciation accompanied by high rates of return. This is in marked contrast to the U.S. pattern."[59] In another 1957 study A. M. Lott of Imperial noted that Canadian pipeline stocks had appreciated more quickly than industrial stocks. He attributed this to the "monopolistic position of Major [sic] Canadian Lines." He added that once these lines are given a charter, "it seems improbable at the present time that 'in the national interest,' there will be any competition for them."[60]

This high profitability continued well into the 1960s. In 1971 a Shell report, "Pipeline Department 1971 Review," commented that "equity

**Table 7.7: After-Tax Rate of Return, Canada's Major Oil Pipelines, 1970 (Per Cent)**

| *Company* | *On Total Investment* | *On Common Equity* |
|---|---|---|
| IPL (Canadian section only) | 15.0 | 24.0 |
| IPL (Consolidated) | 11.1 | 24.5 |
| Trans Mountain | 19.5 | 29.0 |

Source: M. B. McKenzie, "National Energy Board's Review of Oil Pipe Line Returns," Shell Oil Company, March 9, 1972, as cited in Canada, Director of Investigation and Research, Combines Investigations Act, *The State of Competition in the Canadian Petroleum Industry, The Production Sector* (7 vols., Minister of Supply and Services Canada, Ottawa, 1981), vol. 4, p. 167.

investment in Canadian oil pipelines yielded on the average a high rate of return with a relatively low risk attached." The report added:

> Pipelines normally have an almost guaranteed income over a considerably long period. This is inherent in their near-monopolistic position as carriers of oil from proven sources of supply to locations with a relatively secure demand.[61]

Similarly, a 1972 document by Imperial noted that investment in pipelines had been "phenomenally successful," attributing that success to the permit system which gives the lines "a virtual monopoly for Canadian crude deliveries."[62]

Another indication of the profitability of the lines was a 1972 Shell table, which listed the rates of return of several pipeline companies in 1970. The returns for IPL and Trans Mountain are listed in Table 7.7. These rates should be compared to what the National Energy Board (NEB) and the oil industry itself considered as adequate.

The NEB derives its rate of return estimate by dividing fully depreciated total investment, *i.e.*, both debt and equity investment, into the sum of after-tax profits plus interest. In 1970 the NEB and the oil industry accepted a 9 per cent rate as more than adequate for gas pipelines.[63] Since oil lines are inherently no more risky than gas lines, this should also be the rate for oil lines. As can be seen in Table 7.7, both lines earned considerably above this rate. Their after-tax rates of return on common equity were considerably higher, and well above the average for all industries.

These high profits, which presumably would be used in Bechtel's "normal financial approach," had two implications for Canada. First, consumers undoubtedly paid a higher price for petroleum products than they would have paid had pipeline profits been lower. Second, the estimated delivered price of Alberta crude in Montreal was higher than that of Venezuelan crude. That the greater Alberta costs could be attributed to the high pipeline charges can be inferred from a report by Walter J. Levy, who was hired by the independent producers to examine the impact of a pipeline.[64] Levy estimated that Venezuelan oil in Montreal would enjoy a 25 cents per barrel advantage over Canadian oil.[65] Included in Levy's estimate of the delivered cost of Alberta oil to Montreal was an 82 cents per barrel pipeline charge.[66] This was 33 cents higher than the charge which Bechtel had estimated was necessary to build a government-owned pipeline which would assure the bondholders a 5 per cent rate of return. Assuming that both Levy's and Bechtel's estimates were correct, then a government-owned pipeline would have given Albertan oil an 8 cents a barrel advantage over Venezuelan. A government-owned line, paying 5 per cent at that time to investors, would not have been a subsidized line. Five per cent was considered then by the oil companies to be a more than adequate rate of return; according to documents unearthed by the Bertrand Commission, IPL regarded 3 per cent as adequate.[67]

The high cost estimates of Alberta oil in Montreal unquestionably contributed to the scuttling of the line. The opponents of the line, among them all the international oil companies with producing properties in Canada (including those without any refining facilities in Montreal),[68] argued that Alberta oil would not be competitive with foreign oil in Montreal. They were aided in their arguments by Walter Levy, the man whom the independents had hired to investigate the economics of the pipeline.

Nowhere in his report did Levy recommend the construction of such a line.[69] The report, in fact, specifically said that Montreal was not a suitable market for Alberta's oil. It stated:

> In sum, the Montreal market is not an obvious direction of expansion from a logistic point of view and it poses the more difficult problems even in the context of competitive price relationships.[70]

### EXPORTS

As an alternative, Levy suggested that Alberta again seek export markets. He wrote:

> If the possibilities of exports to the United States appear to be adequate to the future development of the Canadian oil economy, and the uncertainties attaching to market expansion in the United States are not too discouraging, then Canadian oil may reasonably await a future expansion of its export markets without actively seeking an outlet in eastern Canada.[71]

The Borden Commission summarized Levy's views as follows:

> He felt that the Montreal refiners might be induced or encouraged to arrange for larger volumes of Canadian crude going to the West Coast of the United States as a quid pro quo for the importation of foreign crude by these Montreal refiners. . . . He also felt that in order that there might be an overall agreement between Canada and the United States on a common oil market, it would be necessary that there be a common policy for Canada, the United States and other Western Hemisphere countries. . . .[72]

His views were echoed by the international oil companies. They also proposed increased exports as an alternative to the pipeline and urged the Canadian government "to press the United States Government for exemption of Canadian crude from import controls in order to facilitate the expansion of export markets for Canadian crude in the United States."[73] That Levy's views were similar to those of the international oil companies was not surprising, for Levy, as has been mentioned elsewhere in this book, was a consultant to most of the major oil companies and to the U.S. State Department.

The Borden Commission accepted the international oil companies' contention that Alberta oil was uneconomical in Montreal. Stating that "low cost energy has been and will remain a vital factor in the Canadian economy,"[74] it recommended:

> that no Government action should at this time be taken to ensure the construction of pipe line facilities to transport Canadian crude oil to the Montreal refinery area and that before any such action is taken an opportunity be given to the oil industry to demonstrate that it can find markets in Canada and the United States sufficient to sustain a healthy and vigorous Canadian oil industry with the incentive for further exploration and development.[75]

The Borden Commission thus told the international oil companies and the United States government that unless they opened up markets in the U.S. for Canadian oil the pipeline would be built. It was this pressure that forced the United States to exempt Canada from the restrictions in its import control program. Canadian imports were, in theory, free to enter the United States in unlimited quantities.

In actuality this freedom was restricted by a hidden tariff imposed on them through the system of allocating U.S. refiners' import quotas. As already mentioned, these quotas were granted to all refiners, including inland refiners who were not in a position to import oil. These inland refiners exchanged their quotas with coastal refiners who, in return, provided domestic crude at a discount. These quotas were thus quite valuable to the inland refiners.

The quotas to these refiners, who had no historical record of imports, were based on their refinery runs. Generally, the higher their output, the larger the quota they received, though the larger refiners received a smaller quota for each additional barrel produced. Imports of Canadian crude, however, were not counted as part of a refiner's output.[76] Therefore, unless Canadian crude were priced considerably below U.S. crude, an inland refiner would have no incentive to use Canadian crude. The value of the lost import quota, like a tariff, gave domestic crude a price advantage over Canadian imports.

This constraint was less severe as far as the internationals were concerned. First, many of them had received historical quotas, based on their past imports. Their quotas, therefore, were not dependent on current refinery runs, and switching to Canadian crude involved no loss of quota. Second, those not on a historical quota had high refinery outputs, so that the additional quota they received for each incremental barrel produced was relatively small. Switching to Canadian crude therefore involved a relatively small loss.

This hidden tariff actually served the interests of the internationals. The Canadian government was putting pressure on them to increase exports. Faced with the alternative of a pipeline to Montreal, they had to bow to that pressure. But they did not want the exports to fall into the hands of the independent refiners in the U.S., since the Canadian exemption opened up the possibility that independents in the northern states could receive Canadian crude at a lower cost than domestic. The refusal to count Canadian crude inputs as part of a refinery run served to prevent this. It assured the internationals control of Canadian exports.

### ALBERTAN VS. VENEZUELAN OIL

The internationals, nevertheless, were not too happy about being forced to increase their purchases of Alberta crude. The reason was Alberta's prorationing system. Based on Texas's system, it had been requested by Imperial and the international oil companies operating in Alberta as a means of controlling production and price.[77] They generally regarded prorationing as a means of increasing their profitability. A 1971 memorandum by G. S. Bevan, Western Marketing Regional Manager of Shell, pointed out, for instance, that all government regulation may not be undesirable. "Although arguments for and against this assumption [government control is likely to be less profitable than otherwise] can be marshalled," he wrote, "we have several examples to the contrary in North America: Texas and Alberta Oil [sic] prorationing . . ."[78]

Nevertheless, they were dissatisfied with the way Alberta's prorationing system was administered. Imperial felt that it encouraged entry into production.[79] In their testimony before the Borden Commission, the internationals complained that they owned most of the "shut-in" capacity in Alberta.[80]

The problem was the allocation to each well in the province of an "economic allowance." This allowance granted to each well a production quota large enough to ensure recovery of drilling, completion, and operating costs. Anyone willing to drill a well received such an allowance.[81] This inducement to entry was what worried Imperial. The sum total of the allowances granted to a large number of wells with low productivity almost equalled the total production which the old Oil and Gas Conservation Board permitted. As a result, the output of the highly prolific wells, which were owned mainly by the internationals, was severely restricted.[82] It was this situation that explained the reluctance of the internationals to purchase Alberta oil and about which they complained to the Borden Commission.

Initially the government of Alberta defended the "economic allowance,"[83] which protected the small, independent producers, who formed an important political constituency in the province. But later, in 1965, the government agreed to change its allocation methods. It substituted a "floor allowance" for the "economic allowance." The "floor allowance" covers completion and operating but not drilling costs.[84] This change discouraged entry since there was no longer any guarantee that a new well would recover all its costs. It also gave the more productive wells a higher share of total demand since the sum total of the floor allowances represented a smaller portion of total demand than the sum total of the economic allowances.[85] The internationals were thus able to obtain a larger share of Alberta production and became far more willing to purchase Alberta oil.

Nevertheless, they did not want to use that oil in Montreal for the simple reason that Venezuelan oil was more profitable to the parents of the Canadian subsidiaries. First, Venezuelan oil was not prorationed. The internationals could therefore supply Montreal entirely with their own oil. It was not unnatural that the internationals would have what Levy called "a commercial preference"[86] for their own product. This "commercial preference" still existed even after Alberta changed its allocation system in a way to grant the internationals higher allowables. In Venezuela the internationals could lift 100 per cent of their own oil. Even the more generous prorationing changes in Alberta precluded them from lifting 100 per cent.

A second factor was the fear that the Venezuelan government might nationalize its oil fields. The internationals wanted to deplete the Venezuelan oil fields as quickly as possible in order to maximize their take from Venezuela before nationalization. They therefore discounted their Venezuelan oil heavily, *i.e.*, lowered the implicit or "shadow" price that they were charging themselves for their Venezuelan crude. While this price represented nothing more than an internal accounting value, it did dictate the sources to which their refineries should turn for oil. Given the heavy internal discount on Venezuelan oil, it was quite rational for the internationals' refineries to prefer that oil. A pipeline to Montreal, however, would have forced the internationals to substitute Albertan for Venezuelan oil, thereby interfering with the rapid depletion of Venezuelan reserves.

Another reason for opposing the pipeline involved the ownership links between the internationals and their subsidiaries in Canada and in Venezuela. Standard Oil (New Jersey), for instance, owned approximately 70 per cent of the shares of Imperial Oil and 95 per cent of the shares in Creole, its subsidiary in Venezuela.[87] This meant that Jersey had claims to 95 per cent of the profits of Creole as opposed to only 70 per cent of the profits of Imperial. It thus had a vested interest in siphoning off profits from the latter to the former. One way Jersey did this was to charge Imperial the posted price for Creole's Venezuelan crude. All the internationals, in fact, made their Canadian affiliates pay the posted price for imported crude.[88] These posted prices were above "fair market value,"[89] *i.e.*, the prices at which the internationals sold their oil to non-affiliates. In 1960, for instance, Imperial was paying Creole the posted price of $2.22 per barrel while, at the same time, Creole was selling its oil to non-affiliates at $1.82 a barrel—a discount of 40 cents per barrel.[90]

An internal Imperial memo, written in 1960, revealed that, even if Imperial did not pay the posted price for its Venezuelan crude imports, it would have made a greater profit by increasing its Alberta production to supply Montreal than Creole would have made by increasing its Venezuelan production by the same amount. Imperial would, according to its own estimates, have earned an after-tax profit of $1.41 per barrel on each additional barrel produced. This profit, it might be added, was derived from an average well-head price of $2.50, indicating that the cost of Alberta oil may have been considerably less than the cost estimates given by the oil companies to the Borden Commission.

As can be seen in Table 7.8, this $1.41 per barrel profit was considerably greater than the profit Creole would have received had Imperial paid the posted price to Creole, not to mention the discounted

**Table 7.8: After-Tax Profits to Imperial, Creole, and Standard Oil (New Jersey) on Each Barrel of Incremental Crude Supplied to Montreal, 1960**

| | *Imperial* | *Creole* | |
|---|---|---|---|
| | | *Posted* | *Discounted* |
| Average Well-Head Price | $2.50 | $2.22 | $1.82 |
| After-Tax Profit to Subsidiaries | 1.41 | 0.90 | 0.69 |
| Profits Going to Jersey | 0.99 (70%) | 0.85 (95%) | 0.65 (95%) |
| Income Tax Paid by Jersey | -0.22 | — | — |
| After-Tax Profit to Jersey | 0.77 | 0.85 | 0.65 |

Source: Canada, Director of Investigation and Research, Combines Investigation Act, *The State of Competition in the Canadian Petroleum Industry, The Domestic Sector: An Overview of the Environment, Industry Behaviour and Performance* (7 vols., Minister of Supply and Services Canada, Ottawa, 1981), vol. 2., p. 12.

price. Economic rationality would have dictated that Montreal be provided with Canadian oil. The interests of Jersey, however, dictated otherwise. Because of its greater financial control of Creole and because of the U.S. income taxes it had to pay on its share of Imperial's profits, Jersey found that it was more profitable to supply Montreal with Creole's crude at the posted price of $2.22 per barrel. The construction of a pipeline obviously did not serve Jersey's interests, even though it apparently would have served the interests of its Canadian subsidiary, Imperial Oil. Needless to say, in this case, as in so many others, the interests of the parent prevailed over those of its affiliates.

The internationals were backed in their opposition to the line by the U.S. government. This backing stemmed not only from the U.S.'s traditional policy of rendering support to its oil companies abroad, but also out of special concern the impact the pipeline could have on the economy of Venezuela. Because Canada at that time was Venezuela's second best customer, accounting for 16 per cent of the latter's export sales,[91] a significant cut in Canadian imports of Venezuelan oil would have inflicted serious damage on the Venezuelan economy, especially when that cutback would be added to the effects of the U.S.'s own import restrictions. The United States feared the political consequences of a major economic downturn in a country in which U.S. corporations had invested heavily and which was located in an area which has traditionally been under U.S. domination.[92]

The U.S. was actually looking to Canada to bail it out of the problems caused by its policy of limiting imports. The U.S. government and the international oil companies apparently assured the Venezuelans that some of their lost markets in the U.S. would be recovered in the fast-growing Montreal market. The failure of Canada to build a pipeline made the U.S. import program more acceptable to Venezuela.

## The National Oil Policy

For whatever reason, the Borden Commission accepted the position of Levy, the international oil companies, and the U.S. government by abandoning a Montreal pipeline in favour of the National Oil Policy (NOP). Under the NOP, which was adopted in 1961 by the Canadian government,[93] the Montreal area and the Maritime provinces were left wholly dependent on imported oil. The rest of Canada, *i.e.*, the entire area west of the Ottawa River, was reserved for domestic crude. At the same time, exports to the U.S., as already mentioned, were to be increased.

This cartel-like division of the market corresponded to the interests of the international oil companies. Estimates of transportation costs between Toronto and Montreal have, for instance, been at between 15 and 25 cents per barrel.[94] Given Jersey's desire to import Venezuelan oil at the posted price to Montreal, the point at which transportation costs equalized the difference between well-head prices of domestic and imported crude was somewhere between Toronto and Montreal. This is precisely where the NOP placed the line. By placing it at that point, the NOP enabled Jersey to perpetuate that division of the Canadian market that corresponded to Jersey's interests. As the Bertrand Report commented in 1971:

> The National Oil Policy, therefore, when announced in 1961, did little more than recognize the supply pattern that had been developed by the leading firm—Exxon's Canadian subsidiary. . . the National Oil Policy chose not to alter drastically patterns that had developed in the post-war era.[95]

The NOP did serve to increase exports to the United States. As Table 7.9 shows, exports rose dramatically in the years following the Borden Commission report in 1959 and the institution of the NOP in 1961, reaching their peak in 1973. What is more significant is that they grew at a much faster rate than domestic production. Between 1958 and 1973, exports increased by over 1300 per cent. Production, on the other hand, grew by less than 300 per cent. In 1958 less than one-fifth of production was exported; by 1973, two-thirds of production was exported.

What had been dubbed a "National Oil Policy" really turned into a "Continental Oil Policy," with Canadian oil production becoming in-

**Table 7.9: Canada's Crude Oil Exports and Production, Selected Years, 1958–73**

| *Year* | *Exports* (Millions of Barrels) | *Production* (Millions of Barrels) | *Exports as Percentage of Production* *(%)* |
|---|---|---|---|
| 1958 | 29.7 | 164.7 | 18.0 |
| 1959 | 33.7 | 183.6 | 18.4 |
| 1960 | 41.3 | 189.5 | 21.9 |
| 1961 | 67.3 | 220.8 | 30.5 |
| 1965 | 107.9 | 291.9 | 37.0 |
| 1969 | 202.7 | 399.9 | 50.7 |
| 1970 | 244.5 | 448.3 | 54.5 |
| 1971 | 274.0 | 476.1 | 57.6 |
| 1972 | 340.9 | 541.6 | 62.9 |
| 1973 | 420.2 | 635.9 | 66.1 |
| Per Cent Change | | | |
| 1958–73 | 1,314.8 | 286.1 | |

Calculated from: Canadian Petroleum Association, *Statistical Year Book*, various years.

creasingly dependent on U.S. markets. President Nixon tried to take advantage of this growing dependence when, in 1970, he attempted to force Canada to agree to a "common or harmonized energy policy with the U.S."[96] The U.S. demanded that Canada make "provision for limiting or offsetting Canadian vulnerability to an interruption of its own imports" as a "precondition to unrestricted entry" into the U.S.[97] Nixon thus unilaterally changed the conditions for granting Canada the unrestricted entry that had been granted in return for vetoing the Montreal pipeline.

The U.S. had a reason for its change in attitude. By 1970 the world oil picture had changed dramatically. Instead of the surplus that characterized the late 1950s and most of the 1960s, shortages began to appear. Within the United States, surplus capacity was fast disappearing. The East Coasts of both Canada and the United States became vulnerable to a supply cut-off. Interestingly enough this question of Montreal's being subjected to an interruption in supplies had been raised at the Borden Commission's hearings. The oil companies dismissed this danger on the grounds that "the matter of national security. . . is more complex than had been indicated by the proponents"[98] of the pipeline. They also argued that "if the entry of Canadian crude into the Montreal market had the effect of reducing the attractiveness of the industry for investment, as well might be the case, a weaker industry would result to the detriment of Canada's ultimate security."[99] Similar arguments, as we shall see in the next chapter, were used in Europe to induce that part of the world to shift to imported oil. The results proved disastrous for Europe and had serious repercussions for Canada.

By 1970 these arguments were no longer valid. The Canadian vulnerability, which resulted directly from U.S. policy, also threatened the United States. The U.S. feared that if imports into eastern Canada were suddenly cut off, Canada would either turn to the U.S. or compete with the U.S. in other oil-producing areas to replace these imports, thereby reducing "the security value of U.S. imports from Western Canada."[100]

To restore the "security value" of Canadian oil, the U.S. wanted Canada to increase storage facilities in Montreal and other eastern centres from a sixty- to a ninety-day capacity, at a cost to Canada of approximately $100 million.[101] The U.S. demanded also, as a condition for granting Canada a new exemption from import restrictions, that American companies gain free "access to natural gas."[102] Unlike with oil, neither Alberta nor Ottawa was anxious at that time to increase exports of natural gas, and both levels of government had imposed restrictions on such exports. Nixon wanted these restrictions removed as part of a "common or harmonized" energy policy.

Nixon was unable to force Canada to agree to this "energy" policy. While there was unquestionably a *de facto* common oil policy in existence at that time, it was not extended into a common energy policy, which would include natural gas, coal, electricity, and atomic energy as well as oil. Nixon was unable to enforce his will because of the rapid depletion of U.S. oil reserves. The U.S., by the early 1970s, found itself in desperate need of oil imports. It was in no position to exclude oil from Canada or anywhere else. It thus abandoned its import control program.

Canada also abandoned its NOP at about the same time. The shortage of oil in Montreal during the 1973/74 oil crisis demonstrated to all Canadians its fundamental weakness: the danger of having an important section of the Canadian economy rely completely on imported oil. It had also prevented Canada from having one integrated oil market, dividing the country instead into two markets, one served by foreign and the other by domestic oil. Given the changed world situation, the U.S. did not object too strenuously to the abandonment of the NOP.

While it is questionable how well the NOP served Canada's interests, there was no question that it had indeed served the interests of the United States and the international oil companies.

The international oil companies benefited not only because they could sell their own oil, which was not prorationed, in the Canadian market but also because they could increase their sales of Middle Eastern oil to Canada. The U.S. control program restricted their scope for selling this oil in the U.S. Faced with a mounting surplus of Middle East oil and pressures by the host governments to market that oil, they did not want Canada to impose restrictions similar to those of the United States. The adoption of the NOP enabled them to shift some Middle East oil from the closed American to the open Canadian market. As can be seen in Table 7.10, Canadian imports of Middle East oil grew much more rapidly than those from Venezuela, despite increases in Canadian imports of Venezuelan oil brought about by the U.S. pressures mentioned earlier (see page 158). While this increased dependence on Eastern Hemisphere oil may not have been in conformance with State Department policy, it nevertheless enabled the U.S.-owned international oil companies to dispose of some of their surplus Middle East oil, a surplus that had been exacerbated by the U.S. oil import program. The U.S. government presumably preferred this increased Canadian dependence on Middle Eastern oil to the building of the Montreal pipeline and the concomitant adoption of import controls. However, these increased Canadian imports of Middle East oil were insufficient to deal with the Middle East surplus. After all, the Canadian market potential, consisting of fewer than 20 million people, was small when compared to that of Western Europe. It was only natural, therefore, that the international companies would look to this area as the main market for their growing surplus of Middle East oil. In this they were actively supported by the U.S. government.

**Table 7.10: Canadian Imports of Crude Oil from Venezuela and the Middle East, 1958–65 (Thousand Barrels per Day)**

| *Year* | *Venezuela* | *Middle East* |
|---|---|---|
| 1958 | 197 | 87 |
| 1959 | 204 | 105 |
| 1960 | 200 | 138 |
| 1961 | 225 | 132 |
| 1962 | 236 | 124 |
| 1963 | 249 | 143 |
| 1964 | 282 | 98 |
| 1965 | 236 | 147 |
| Per Cent Change, 1958–65 | 20 | 69 |

Source: Canadian Petroleum Association, *Statistical Year Book*, various years.

## Notes

1. E. Gray, *The Impact of Oil* (The Ryerson Press/Maclean-Hunter Ltd., Toronto, 1969), p. 14.
2. Ibid, p. 15.
3. E. Gray, *The Great Canadian Oil Patch* (Maclean-Hunter, Toronto, 1970), p. 248.
4. Ibid., pp. 108–12.
5. Ibid., p. 126.
6. Ibid., p. 134.
7. Ibid., pp. 134–35.
8. A. Plotnick, *Petroleum: Canadian Markets and United States Foreign Trade Policy* (University of Washington Press, Seattle, 1964), p. 15.
9. Alberta first instituted market demand prorationing in 1950. See S. Thompson, "Prorationing of Oil in Alberta and Some Economic Implications," unpublished M.A. thesis, University of Alberta, 1968, p. 24.
10. E. Gray, *The Great Canadian Oil Patch*, p. 136.
11. Ibid.
12. The Superior line was 200 kilometres shorter than a Thunder Bay line would have been. Imperial Oil estimated that an all-Canadian line would have added $10 million to construction costs and $1 million per year to operations costs. See ibid.
13. Ibid, pp. 137–38.
14. Ibid, p. 140.
15. Ibid.
16. Ibid., pp. 141–42.
17. Ibid., pp. 149–50.
18. Canada, Royal Commission on Energy, *Second Report*, (Ottawa, 1959), p. 28.
19. Ibid.
20. Ibid.
21. Ibid.
22. Ibid., p. 29.
23. Ibid., p. 30.
24. E. Gray, *The Great Canadian Oil Patch*, p. 145.
25. Ibid.
26. *Edmonton Journal*, August 21, 1982, and *Globe and Mail*, July 14, 1982.
27. Canada, Royal Commission on Energy, *Second Report*, p. 30.
28. Ibid., p. 31.
29. Ibid., p. 32.
30. Ibid., pp. 44–45.
31. Ibid., p. 45.
32. Ibid.
33. E. Shaffer, *The Oil Import Program of the United States, An Evaluation* (Praeger, New York, 1968), p. 109.
34. Canada, Royal Commission on Energy, *Second Report*, p. 28.

35. Canada, Director of Investigation and Research, Combines Investigation Act, *The State of Competition in the Canadian Petroleum Industry, The Domestic Sector: An Overview of the Environment, Industry, Behaviour and Performance* (7 vols., Minister of Supply and Services Canada, Ottawa, 1981), vol. 2, p. 5.
36. Canada, Royal Commission on Energy, *Second Report*, p. 92.
37. Ibid., p. 95.
38. Ibid., p. 99.
39. Ibid., p. 98.
40. Ibid., p. 36.
41. Ibid., p. 100.
42. P. Smith, *The Treasure-Seekers, The Men Who Built Home Oil* (Macmillan of Canada, Toronto, 1978), p. 194.
43. Ibid.
44. Canada, Royal Commission on Energy, *Second Report*, p. 100.
45. D. Crane, *Controlling Interest, The Canadian Gas and Oil Stakes* (McClelland and Stewart, Toronto, 1982), p. 53.
46. Canada, Royal Commission on Energy, *Second Report*, pp. 101–2.
47. Ibid., p. 106.
48. Ibid., p. 104.
49. Smith, *The Treasure-Seekers*, p. 200.
50. Canada, Royal Commission on Energy, *Second Report*, p. 106.
51. Ibid., p. 107.
52. Ibid.
53. J. K. Whiteford, Gulf's president, adamantly opposed the sale of Gulf gas to a government-owned entity. See Crane, *Controlling Interest*, p. 51.
54. Ibid.
55. Ibid., p. 52.
56. Ibid.
57. Canada, Royal Commission on Energy, *Second Report*, p. 104.
58. Ibid., p. 107.
59. Canada, Director of Investigation and Research, Combines Investigation Act, *The State of Competition in The Canadian Petroleum Industry, The Production Sector*, vol. 4, p. 163.
60. Ibid, p. 166.
61. Ibid.
62. Ibid.
63. Ibid., p. 167.
64. P. Smith, *The Treasure-Seekers*, p. 194.
65. Ibid., p. 195.
66. Ibid., p. 200.
67. Canada, Director of Investigation and Research, Combines Investigation Act, *The State of Competition in the Canadian Petroleum Industry, The Production Sector*, vol. 4, p. 161.
68. Canada, Royal Commission on Energy, *Second Report*, p. 113.
69. P. Smith, *The Treasure-Seekers*, p. 195.
70. Canada, Royal Commission on Energy, *Second Report*, p. 112.
71. Ibid.

72. Ibid.
73. Ibid., p. 114.
74. Ibid., p. 145.
75. Ibid., p. 144.
76. E. Shaffer, *The Oil Import Program of the United States*, p. 163.
77. Canada, Director of Investigation and Research, Combines Investigation Act, *The State of Competition in the Canadian Petroleum Industry, The Domestic Sector*, vol. 2, p. 9.
78. Ibid, p. 1.
79. Ibid, p. 9.
80. Canada, Royal Commission on Energy, *Second Report*, p. 115.
81. S. Thompson, "Prorationing of Oil in Alberta and Some Economic Implications," unpublished M.A. thesis, University of Alberta, 1968, pp. 26–27.
82. Ibid, p. 34.
83. Canada, Royal Commission on Energy, *Second Report*, p. 115.
84. Thompson, "Prorationing of Oil in Alberta and Some Economic Implications," p. 42.
85. Ibid.
86. Smith, *The Treasure-Seekers*, p. 196.
87. Canada, Director of Investigation and Research, Combines Investigation Act, *The State of Competition in the Canadian Petroleum Industry, The Domestic Sector*, vol. 2, p. 12.
88. Ibid., *Findings, Issues and Remedies*, vol. 1, p. 63, and Canada, Royal Commission on Energy, *Second Report*, p. 114.
89. Canada, Director of Investigation and Research, Combines Investigation Act, *The State of Competition in the Canadian Petroleum Industry, Findings, Issues and Remedies*, vol. 1, p. 63.
90. Ibid., *The Domestic Sector*, vol. 2, p. 12.
91. *Activitad Petrolera en 1962: Separate del Informe Economico del Banco Central de Venezuela Correspondiente al Bjercicio Annual 1962, Capitulo de Hidrocarbons* (Banco Central de Venezuela, Caracas, n.d.), pp. 9–29.
92. The Canadian government, incidentally, expressed a similar concern when, in 1969, a delegation from the Independent Petroleum Association of Canada (IPAC), the organization representing the independent producers, presented a brief to the cabinet urging that the pipeline to Montreal be built. The cabinet refused this request. According to the minutes of the meeting at which the IPAC delegation reported to the organization's executive committee, the cabinet stated that "Canada, along with the U.S., must protect Venezuela's interest and ensure that this country remained within the western hemisphere community of democratic nations." (Smith, *The Treasure-Seekers*, p. 207.)
93. Canada, Director of Investigation and Research, Combines Investigation Act, *The State of Competition in the Canadian Petroleum Industry, The Domestic Sector*, vol. 2, p. 11.
94. Ibid.
95. Ibid, p. 13.

96. U.S. Cabinet Task Force on Oil Import Control, *The Oil Import Question: A Report on the Relationship of Oil Imports to National Security* (U.S. Government Printing Office, Washington, 1970), p. 94.
97. Ibid.
98. Canada, Royal Commission on Energy, *Second Report*, p. 118.
99. Ibid.
100. U.S. Cabinet Task Force on Oil Import Control, *The Oil Import Question*, p. 94.
101. *Financial Post*, June 19, 1971.
102. U.S. Cabinet Task Force on Oil Control, *The Oil Import Question*, p. 94.

CHAPTER EIGHT

# The Control of World Oil: The Transformation of Europe

As has already been noted, the U.S. oil companies had been actively seeking markets in Europe since the end of World War II. The U.S. government, through the Marshall Plan and other aid programs, supported them fully in their efforts to penetrate the Continent. This support was the key to U.S. postwar oil policy.

Though in line with the general policy of encouraging U.S. business expansion abroad and with the specific policy of finding markets for U.S. oil companies to enable them to keep their Middle East concessions, this support had a more basic aim—the permanent control of Europe. In the immediate postwar years, there was a fear in Washington that several European countries, mainly France and Italy, might go communist. Moreover, in such countries as Britain and West Germany, communist-led unions were dominant in the coal fields.[1] Coal, which was then Europe's principal energy source, might thus come under Communist control.

Middle East oil, on the other hand, seemed at that time to be firmly under U.S. control. By inducing Europe to switch from coal to oil, the U.S. could keep Europe within its orbit. This strategy was openly bandied about by U.S. policy-makers in the early postwar years. In his diary, the then U.S. secretary of defense, James Forrestal, recorded a conversation that he had with Senator Owen Brewster, chairman of the Senate Commerce Committee, on this subject.

> Brewster said that he had had a long talk with John D. Rockefeller, Jr.... Europe in the next ten years may shift from a coal to an oil economy, and therefore whoever sits on the valve of Middle East oil may control the destiny of Europe.[2]

The control of Europe was fundamental for U.S. strategy. Though initially motivated by the desire to keep Europe non-Communist, it later became necessary to ensure that Europe remained under U.S. hegemony. What the U.S. feared was that an economically independent Europe would pose a serious competitive threat to U.S. enterprises. It was not in

**Table 8.1: Fossil Fuels[a] Consumed in OECD Europe, Selected Post-World War II Years (Per Cent)**

| | Type of Fossil Fuel | | |
|---|---|---|---|
| *Year* | *Solid*[b] | *Liquid*[c] | *Gas*[d] |
| 1950 | 85 | 15 | x |
| 1955 | 77 | 22 | 1 |
| 1960 | 64 | 34 | 2 |
| 1965 | 47 | 51 | 2 |
| 1970 | 30 | 63 | 7 |
| 1973 | 23 | 66 | 11 |
| 1974 | 24 | 63 | 13 |
| 1978 | 22 | 63 | 15 |

a. The fossil fuels are the hydrocarbons, *i.e.*, coal, oil, natural gas, and other fuels created by the photosynthesis process. They are, in fact, stored solar energy. Excluded from this table are such non-fossil fuel energy sources as: hydroelectric power, atomic and tidal energy. These latter energy sources constitute only a small fraction of OECD Europe's supply. In 1980 the European members of OECD (Organisation for Economic Cooperation and Development) were: Austria, Belgium, Denmark, Finland, France, Federal Republic of Germany, Greece, Iceland, Ireland, Italy, Luxembourg, the Netherlands, Norway, Portugal, Spain, Sweden, Turkey, and the United Kingdom. Fuel for overseas marine bunkers is included in the consumption totals.
b. The solid fuels are: coal, coke, lignite, peat, wood, and the like.
c. The liquid fuels are: crude oil, petroleum products, and natural gas liquids.
d. The gas fuels are: natural gas and town gas.

Sources: 1950, 1955—Organisation for Economic Cooperation and Development, *Energy Policy: Problems and Objectives* (OECD, Paris, 1966), p. 150; 1960, 1965, 1970, 1973—OECD, *Energy Balances of OECD Countries, 1960*/1974 (OECD, Paris, 1976), pp. 41, 46, 51, and 54; 1974, 1978—International Energy Agency, *Energy Balances of OECD Countries, 1974/1978* (IEA, Paris, 1980), pp. 7, 11.

the U.S. interests to allow Europe to become strong enough to challenge the U.S. dominance of the world economy. By tying Europe to U.S.-owned Middle East oil, the U.S. could keep Europe in check. Thus, in the early postwar years, both the U.S. government and the U.S. oil companies started a campaign to force Europe to become dependent on oil.

## Europe's Dependence on Coal

Their task was not an easy one. As can be seen in Table 8.1, solid fuels[3] furnished 85 per cent of Western Europe's fossil fuel[4] requirements in

1950; oil furnished the remainder. Furthermore, since automobile ownership in Europe was relatively low, the transportation sector did not provide the main market for oil as it did in the United States. As late as 1960, as Table 8.2 shows, the transportation sector accounted for only 30 per cent of oil consumption in Europe. By comparison, it accounted for approximately 50 per cent in the United States.

To rely on the transportation sector to increase Europe's dependency on oil promised to be a long-run process. Aside from an increase through the conversion of the railroads from coal to diesel locomotives (or alternatively to electric trains, with the electricity being generated by oil-fueled thermal power stations), an increase in oil demand by the transport sector depended mainly on the automobilization of Europe, which implied not merely a very significant expansion of private automobile ownership but also the building of ancillary facilities, like highways, service stations, and garages. Such a development could not occur overnight.

A much more immediate prospect was the conversion of industry, electricity generation, and home heating from coal to oil. Given the proper price incentives, industry could be expected to change its equipment in a relatively short time. That there was considerable scope for

**Table 8.2: Oil Use by Sector, OECD Europe and United States, 1960 (Per Cent)**

| *Sector* | *OECD Europe* | *United States* |
|---|---|---|
| Transportation | 30 | 49 |
| (Road) | (23) | (41) |
| Industry | 24 | 8 |
| (Iron and Steel) | ( 4) | (--) |
| Other[a] | 27 | 28 |
| Electricity Generation | 6 | 3 |
| Marine Bunkers | 5 | 4 |
| Manufacturing of Town Gas | 1 | – |
| Non-Energy[b] | 7 | 8 |

a. Mainly home heating but also includes heating of office buildings, street lighting, and agriculture.
b. Production of bitumen, lubricants, and naphtha. Naphtha is used extensively as a petrochemical feed stock.

Source: OECD, *Energy Balances of OECD Contries, 1960/1974* (OECD, Paris, 1976), pp. 41, 101.

**Table 8.3: Coal's Share of Sectoral Fossil Fuel Energy Requirements, OECD Europe and United States, 1960 (Per Cent)**

| *Sector* | *OECD Europe* | *United States* |
|---|---|---|
| Industry | 64 | 29 |
| Electricity Generation | 86 | 66 |
| Other[a] | 67 | 10 |

a. Mainly home heating but also includes heating of office buildings, street lighting, and agriculture.
Source: OECD, *Energy Balances of OECD Countries, 1960/1974* (OECD, Paris, 1976), pp. 41, 101.

such a transformation can be gleaned from a comparison of coal consumption in these sectors in Western Europe and in the United States in 1960. As Table 8.3 shows, coal furnished a far larger portion of the energy requirements of these sectors in Western Europe. Because residual fuel oil is competitive with coal, it can make inroads into coal's markets by selling at a price lower than that of coal. By charging a relatively low price for residual, the oil companies were able to drive coal from its industrial and electricity-generating markets. As can be seen in Table 8.4, they adopted this pricing pattern in Europe, a pattern similar to that in the United States.

## European Pricing Patterns

The data in Table 8.4, taken from estimates made by the United Nations Economic Commission for Europe (UNECE), show that in the early 1950s the price of gasoline both in Europe and in the United States was approximately $2.50 per barrel higher than the price of residual fuel oil. In both places the price of residual was below that of crude. In other words the pricing policies of the oil companies led to one finished product, residual fuel oil, being sold at a lower price than the raw material. In both places the price of residual was also lower than the price of American coal, the lowest-priced coal on the market, even after including freight costs to Europe.

At first glance, this European price pattern is one that would be expected in a competitive economy. Since Europe had traditionally imported most of its products from the Western Hemisphere, it would seem only natural that the landed prices of these imports would set a

**Table 8.4: Petroleum Product Pricing Patterns, North-Western Europe (1951) and United States[a] (1950) ($ per Barrel)**

| *Product* | *Europe* | *United States* |
|---|---|---|
| Gasoline | $5.19 | $4.20 |
| Residual Fuel Oil | 2.74 | 1.65 |
| Difference | 2.45 | 2.55 |
| Refinery Acquisition Cost of Crude Oil | 3.22 | 2.72 |
| Oil Equivalent Price of U.S. Coal | 5.50[b] | 1.91[c] |
| Refinery Margin[d] | 0.52 | 0.66 |

a. Charged at Gulf of Mexico ports.
b. C.I.F., Europe.
c. F.O.B., United States.
d. The refinery margin is the difference between the revenue a refiner receives from the sale of a barrel of its product and the acquisition cost of a barrel of crude.

Sources: Oil prices and margins—UNECE, *The Price of Oil in Western Europe* (Geneva, 1955), Annex, p. 4; coal prices—Organisation for European Economic Cooperation, *Europe's Growing Needs of Energy—How Can They Be Met?* (OEEC, Paris, 1955), p. 45.

ceiling on the prices European refineries could charge. These similar pricing patterns would thus reflect the integration of the world product market.

Upon closer examination, however, this particular explanation does not hold water. First, by the early 1950s, Europe was already becoming a major refining centre. Product imports, and especially those from the Western Hemisphere, were falling dramatically, thereby loosening the links to the U.S. product market. Second, and even more important, was the discrepancy between refinery margins, *i.e.*, the difference between the revenue a refinery receives from selling the products of a barrel of crude and the cost of acquiring that barrel. As can be seen on Table 8.4, according to the calculations of the UNECE, these margins were higher in the United States. Competition presumably would give rise to forces to equalize these margins. But the price structure prevented this equalization from occurring. This structure did not conform to the differences in product yields between refineries in Europe and in the United States. The refineries in Europe yielded a much higher percentage of residual fuel oil than those in the U.S. and, conversely, a much lower percentage of gasoline. These patterns are shown in Table 8.5.

More than half of the output of European refineries consisted of

**Table 8.5: Petroleum Yield Patterns, North-Western Europe (1951) and United States (1950) (Per Cent)**

| *Product* | *Europe* | *United States* |
|---|---|---|
| Gasoline | 26 | 49 |
| Kerosine | 2 | 10 |
| Distillate Fuel Oil | 20 | 24 |
| Residual Fuel Oil | 51 | 17 |

Source: UNECE, *The Price of Oil in Western Europe* (Geneva, 1955), Annex, p. 4.

residual fuel oil, the lowest-priced product. In the United States, on the other hand, almost half of the output consisted of gasoline, the highest-priced product. Hence the difference in refinery margins.

Under competition, rational behaviour would have dictated that refineries in Europe follow the example of those in the United States by increasing their gasoline yields through the installation of cracking facilities, which would have allowed them to convert a portion of the residual into gasoline. Residual could have thus served as a substitute for the higher-priced crude in the production of gasoline. This use of residual would have significantly raised European refinery margins.

Despite the incentives presented by the price structure, the European refineries, which were mainly owned by the internationals, did not significantly reduce their residual yields. Their production policies were dictated by the consumption patterns in Europe, which differed markedly from those in the United States. These differences are illustrated in Table 8.6.

**Table 8.6: Petroleum Product Consumption Patterns, Western Europe and United States, 1953 (Per Cent)**

| *Product* | *Europe* | *United States* |
|---|---|---|
| Gasoline | 31 | 51 |
| Kerosine | 6 | 5 |
| Distillate Fuel Oil | 25 | 21 |
| Residual Fuel Oil | 39 | 23 |

Source: UNECE, *The Price of Oil in Western Europe* (Geneva, 1955), Annex, p. 3.

Granted that the refineries followed the European consumption pattern, the question arises as to why the European price structure did not also follow this pattern. In its report, the UNECE noted that

> since the demand for fuel oil relative to that of motor spirit is much higher in western Europe than in the United States, the difference between the prices of these products would be much smaller if prices in western Europe were determined in the light of the conditions of demand and production ruling there.[5]

Presumably the refineries could have increased their margins by raising the price of residual, the product for which there was the greatest demand in Europe. But this short-term gain would have endangered the long-term goal of penetrating the coal market.

The experience of the prewar years testified to the importance of this link between fuel oil prices and market penetration. In those years the price of residual was higher than that of European coal, and this relatively high price limited the market for residual. As a result, the few refineries that did exist on the Continent were forced to export their fuel oil.[6] Coal's dominant position remained unshaken.

## European Refinery Expansion

The postwar pricing pattern reversed this situation. Within a short time Europe became a net importer of fuel oil. The expanding European market served as an inducement to enlarge refinery capacity and distribution facilities in the region. As a result, the geographical distribution of U.S. investment in the petroleum industry shifted decisively towards Europe. As Table 8.7 shows, U.S. oil companies had placed slightly over one-tenth of their foreign investments in Europe in 1950. By 1960, this portion had risen to approximately one-sixth. Ten years later, in 1970, it was more than one-quarter. By 1974, it was almost one-half. By this time Europe had become the principal area of U.S. oil investments.

## The Assault on Coal

Given the relatively low margins received by European refineries, this heavy investment in refineries appears irrational. But, in reality, it was a very rational move. It, first of all, enabled the oil companies to increase their sales of Middle East crude. These increased sales satisfied both the demands of the Middle East governments for more revenues and the

**Table 8.7: Geographical Distribution of U.S. Private Direct Foreign Investment in the Petroleum Industry, Selected Post–World War II Years (Per Cent)**

| *Year* | *Europe* | *Latin America* | *Canada* | *Other* |
|---|---|---|---|---|
| 1950 | 12 | 35 | 12 | 41 |
| 1960 | 16 | 27 | 25 | 33 |
| 1970 | 28 | 14 | 22 | 37 |
| 1974 | 46 | 17 | 27 | 11 |
| 1975 | 43 | 13 | 24 | 20 |
| 1976 | 44 | 10 | 25 | 21 |
| 1977 | 44 | 11 | 25 | 20 |
| 1978 | 44 | 11 | 25 | 20 |

Calculated from: *Survey of Current Business*, various issues. Based on year-end book values. Because of rounding, the sum of some rows may not equal 100.

*Note*: The shift to Europe was not entirely due to an increase in investment in refining and marketing facilities. It was also due to investment in North Sea oil. Furthermore, some of the decline in the Latin American and Other shares resulted from participation agreements and nationalizations, particularly in Venezuela and in the Middle East. Nevertheless, the petroleum industry invested far more in refining and marketing in Europe than in any region outside of the United States. For the years 1963–76 the industry invested approximately equal amounts in refining and marketing facilities in the United States and Europe, about $29 billion in each area. See Chase Manhattan Bank, *Capital Investments of the World Petroleum Industry*, various issues.

desire of the oil companies for higher profits. Because of the industry's price structure, profits were made in production—the upstream phase—rather than in refining and distribution—the downstream phases.

It must be remembered that only the independent refineries, *i.e.*, those who did not own any crude-producing properties, paid the full price for imported crude. This price was based on the posted price in the Middle East plus transportation and other costs, and it equalled the refinery acquisition costs shown in Table 8.4 (page 172). The integrated refiners, *i.e.*, those owned by the international oil companies, supposedly acquired crude at the tax-paid costs, consisting mainly of production expenses, tax and royalty payments, plus transport and other charges. However, depending on the firms' exact tax and royalty arrangements and their ability to make use of the U.S. tax credit, the real acquisition cost of crude was probably no more than half the nominal cost. Even though the price of residual fuel oil was less than the reported refinery acquisition cost of crude, the residual price exceeded the integrated refiners' actual acquisition costs. Thus, from the point of view of a fully-integrated company, the increased sales of residual fuel oil turned out to be a profitable operation.

A second reason for refinery investment was that the oil companies probably viewed their theoretical losses from reduced refining margins as investments designed to generate greater future income. To the extent they succeeded in closing down coal mines, they would be able to eliminate the restraints on residual fuel oil prices imposed by coal. Knowing that once coal mines are closed down they are very difficult and costly to reopen, the oil companies presumably believed that they could then raise residual prices without fear of competition from coal. In more technical terms, the abandonment of coal mines would lower the price elasticity of demand for fuel oil, thereby significantly increasing the opportunities for raising its price. The companies' fuel oil pricing policy might therefore be viewed as a form of "predatory pricing," a policy often practised by Rockefeller by which he sustained short-term losses in order to drive out his competitors.

Third, to the extent that the companies could eliminate coal as a competitor, they would make Europe dependent on their oil, thus giving them a high degree of control over the European economy. This objective coincided with the U.S. foreign policy aim of keeping Europe under U.S. control.

*Armand Report*

Many Europeans were aware of the implications of the shift to oil. Responding to their concerns, several governments and international agencies conducted studies of and issued reports on energy policy. One of the first of these studies was the Armand Report of 1955, named after its author, Louis Armand, chairman of the board of the Société Nationale des Chemin de Fer Francais, the French national railroad. This report was commissioned by the Council of Ministers of the Organisation for European Economic Cooperation (OEEC), an agency which was an offspring of the Marshall Plan. In his report Armand expressed concern about the inroads foreign oil was making in Europe. "The most outstanding feature of Europe's situation with regard to oil," he wrote, "is undoubtedly the fact that most supplies have to be imported."[7] He complained that these imports, accounting for 97 per cent of total oil consumption, constituted "a heavy drain on foreign exchange, a substantial proportion being in hard currency (dollars)."[8] He advocated instead the development of indigenous energy resources which, he believed, would also reduce energy costs. "Instead of 'more and more power'," he confidently proclaimed, "the slogan will now have to be 'cheap power above all'."[9]

Atomic energy was to be the generator of this cheap power. He gave vent to his unbounded enthusiasm for this new energy source in the following words:

> The first consequence of the dawn of the atomic era must be to abolish the fear of any power shortage in the future; it will be therefore necessary to abandon the idea that power costs in Europe are linked irretrievably with the price of coal and are therefore bound to rise, as the future cost of atomic energy will certainly be lower than the average prices charged in Europe at present.[10]

He nevertheless recognized that "atomic energy cannot alone cover the anticipated increase in demand during the next ten years or so."[11] He therefore conceded that "a certain proportion of the increased demand for power will have to be met from traditional resources in the next 10 or 20 years."[12] Coal was the chief "traditional resource."

His enthusiasm for coal was surpassed only by his optimism about atomic energy. Though admitting that coal prices would rise in the long run, he was sanguine about the possibilities of reducing coal costs in the short term, *i.e.*, in the interim before its displacement by cheaper atomic power.

> There can be no immediate prospect of a reduction in coal requirements which are, on the contrary, likely to increase. . . . Nor should it be forgotten that new equipment will have an important bearing on prices. In these circumstances, the modernisation plans for certain mines will remain just as necessary for many years to come. . . . No mention has been made of technical developments which . . . might lead to substantial progress or even far-reaching changes (e.g., underground gasification . . .). In this connection there is considerable unused potential in the European coal and gas industries which, based on lower prices, should be put to minimum use during the next few years until such time as atomic power is in a position to play a more important part.[13]

He also placed high hopes on developing Europe's own oil and natural gas resources, based on natural gas discoveries in Italy and in southern France.[14] In addition, he regarded the discoveries in Europe's African colonies, and especially those in the French Sahara, as part of Europe's reserves,[15] an attitude not too dissimilar from that of the United

States government to the reserves owned by U.S. companies. Finally, he advocated locating industrial plants near hydroelectric sites in such places as Norway, Austria, Yugoslavia, and Africa.[16] Presumably he expected the latter area to remain permanently under European domination.

What was significant about Armand's report was not his specific proposals but his advocacy of European control over energy sources. Most important, he did not look to oil as the means of supplying the Continent with cheap energy. He emphasized instead oil's drain on Europe's limited hard currency reserves.

After examining his report, the OEEC Council decided in 1955 to establish a Commission on Energy and set up a working party to explore the ways and means of cooperating in the development of nuclear energy.[17]

### *Hartley Report*

Shortly after its founding, the Commission on Energy asked Sir Harold Hartley of the United Kingdom to prepare a report on Europe's energy needs. Like Armand, Hartley expressed concern about Europe's growing energy imports. But his worries, unlike Armand's, extended beyond the mere question of the balance of payments. He was also worried about security of supply. Writing in early 1956, before the Arab-Israeli war of that year and the advent of Europe's first oil crisis, he commented that "there are inevitable risks in the increasing dependence of Western Europe on outside supplies, particularly when most of them must come from one small area of the world."[18] Though conceding that the "incentive to substitute oil for coal has been increased by the rise in coal prices in comparison to those of oil,"[19] he nevertheless argued that "coal must remain the mainstay of the Western European energy economy."[20]

Far less optimistic than Armand about the potential of nuclear energy, he deplored the adverse effects that "exaggerated statements about nuclear energy"[21] had had on investment in the coal industry. At the same time he conceded that coal could not fulfil all of Europe's energy requirements. He forecast that by 1975 domestic coal production would fill slightly less than one-half of Europe's energy needs. The remainder would be supplied by hydro-power, indigenous crude oil and natural gas, and the importation of nuclear fuels, coal, and crude oil. He expected that imported crude oil alone would meet between one-fifth and one-third of total energy consumption.[22] (In actuality, crude oil imports in 1975 accounted for over one-half of total energy requirements.[23]) Apparently he regarded this level of crude imports as acceptable from the point of view of security of supply, based on conversations he had with officials of the oil companies. He reported:

> The oil companies with whom we had discussions assured us that in view of their present knowledge of oil reserves ample crude would be available to meet the maximum demands of our forecasts (*i.e.*, one-third of energy requirements in 1975). They pointed out that oil is an international industry and that Europe's consumption at present represents 15 per cent of the whole, though this might rise to 20 per cent by 1975.[24]

Despite these optimistic assurances, Hartley did not want oil imports to grow because of anticipated balance of payments difficulties. He forecast that the costs of imports would reach $5 billion by 1975, assuming no increase in 1956 price levels, but he added that the prices of these imports "are certain to rise."[25] The only way to prevent future serious balance of payments difficulties, he argued, was to expand coal production through creating conditions favourable for long-term investments in the industry. These conditions did not exist in 1956. But he was confident "that given favourable conditions for investment and development on a long-term basis, the coal industry could make a substantial addition to its output."[26] "In our opinion," he added, "Western Europe will never regret its investments in the coal industry."[27]

To create favourable conditions for investment in coal, Hartley urged that each country in the OEEC adopt an energy policy. While giving lip-service to competition, he clearly implied that policy measures must be substituted for market-place decisions if Europe were to solve its energy problems.

> We think that competition must be allowed to play its full role in giving the consumer maximum choice between energy sources. However, in order to deal effectively with the urgent problems involved in the supply and demand of energy, each Member country will require an energy policy suited to its own circumstances and its needs and resources. This policy should include some measure of coordination between the different forms of energy.[28]

In this he was echoing the view of Armand, who had urged the OEEC to avoid "a situation in which competition between the various forms of energy acts to the detriment of the community as a whole."[29] It is interesting to note that both Armand and Hartley came to this conclusion despite the oil companies' assurances that Europe had nothing to fear from an increased dependency on imported oil.

*Robinson Report*

Following the acceptance of the Hartley Report, the OEEC in 1956 established an Energy Advisory Commission, a "permanent group of high-level experts" under the chairmanship of Professor Austin Robinson of Cambridge University.[30] The nine members of this commission were to act as "independent experts to the Organisation and not as official representatives of their countries of origin."[31] Shortly after its inception, this commission began its own study of European energy problems.

Completed in 1959, this study proposed solutions radically different from those of Armand and Hartley. Robinson's arguments were based on a presumed "change from a protracted period of shortage of energy in total to a state of overall surplus."[32] If there were a danger of a shortage of any fuel, the report continued, "we would expect coal supplies to be near the upper limit."[33] Because of this new situation, the Robinson Commission was not "unduly alarmed at the existence of a 'gap' in European energy supplies."[34] The increasing reliance on imports should pose no special problems because "there are reasonable prospects that the imports can be financed."[35]

Dismissing the balance of payments concerns of the earlier reports, Robinson looked to the Sahara as a major source of oil imports. Assuming, like Armand, that the Sahara would always remain French, he maintained that these imports would be paid for in French currency and "corresponding exports may be more than normally easy."[36] To the extent that the imports would come from other areas, the foreign exchange drain, he claimed, would be more apparent than real. Without giving any quantitative estimates, he concluded that "the net overseas payment, after allowing for interest and dividends on European holdings in the oil companies concerned, is considerably less than the import costs that appear in the trade returns."[37] Ignoring that the bulk of these interest and dividends would flow to American rather than to European interests, he made an interesting, though probably unwitting, admission, namely, that the ability of Europe to pay for its oil imports hinged on the economic and political control of the oil-producing areas. In other words, only through a policy of oil imperialism would Europe be able to finance these imports.

Robinson furthermore felt that manufactured exports could also alleviate the balance of payments problem. He wrote:

> If Western Europe can maintain its share of world markets for manufactures, the import of the increased proportions of the total supplies of energy that have emerged from our study may reasonably be expected to be within the probable limits of its capacity.[38]

The ability to compete in world markets, whether to hold one's share or to increase it, depends on price effectiveness, which in turn depends on low input costs. The Robinson Commission therefore emphasized the cheapness, rather than the security, of energy supplies.

> When formulating a long-term energy policy, the paramount consideration should, in our view, be a plentiful supply of low-cost energy with a freedom of choice to the consumer. . . . The general interest is best served by placing the least possible obstacle in the way of economic development of the newer and cheaper sources of energy.[39]

This emphasis on "low-cost energy" meant a rejection of Hartley's suggestion that long-term investment in coal be encouraged. Assuming that any expansion of coal production would increase costs, Robinson favoured reducing coal output and concentrating it in those pits "where the prospects of increasing productivity are greatest."[40] The coal actually marketed should sell "at a price that will make its use fully competitive with that of alternative forms of energy."[41] Given the residual fuel oil pricing policies of the oil companies, most European coal could not be profitably sold at a price competitive with that of oil. Hence the adoption of a "low-cost energy" policy spelled the end of an indigenous one based on coal.

The Robinson Commission did not merely reject coal. It also downplayed nuclear energy. Claiming that "an active policy of developing nuclear energy secures a relatively small immediate saving of foreign exchange at the cost of a very large investment of indigenous European resources," it suggested that in view of the energy surplus the "Member countries review their plans for the development of nuclear energy."[42] In summary, the Robinson Commission advocated the abandonment of indigenous energy sources in favour of reliance on imported oil.

Interestingly, the commission's report was written after the oil crisis caused by the Arab-Israeli war of 1956–57. Nevertheless, the Commission was not worried about security of supply. In fact the Commission stated quite categorically that "it does not seem likely that shortages of oil supplies will make themselves felt in any acute form by 1975."[43] These assurances of future oil supplies were based on information supplied by the international oil companies, whom the commission had consulted in preparing its report.[44]

Because of their implications for the domestic coal industries, Robinson's recommendations received a mixed reception. In his foreword, the secretary-general of the OEEC was careful to note that "the views expressed and the conclusions reached in this report represent the opinion of this group of independent experts and thus do not necessarily reflect the views of the Organisation as a whole."[45] Rather than formally endorse it, the OEEC Council voted merely to publish it.[46]

Nevertheless, the governments of most OEEC countries followed the recommendations in varying degrees. So did the European Coal and Steel Community. Under rationalization schemes, coal mines were shut down throughout Europe; Belgium and Holland stopped coal production completely; in Germany and France it was reduced drastically.[47] European coal production peaked in 1957 and has declined steadily since that year.[48] As can be seen in Table 8.8, it had fallen by more than 40 per cent by 1973 while total European energy requirements had more than doubled. Coal production in the relatively prosperous year of 1973 was lower than in the Depression year of 1937 and the early postwar year of 1950.

Coal was displaced by oil, and to a lesser extent by natural gas, in thermal power plants, industrial enterprises, office buildings, and households. In all industrial establishments coal consumption fell from 116 mtoe (millions of metric tons of oil equivalent) in 1960 to 79 mtoe in 1973. As Table 8.9 shows, this fall was far more pronounced in industries other than iron and steel. These latter industries must consume coke in the production processes, and they therefore had little room to substitute oil for coal. Other industries, which used coal mainly for heating purposes, were able to substitute residual fuel oil. Thus, while

**Table 8.8: Indigenous Solid Fuels Production and Total Energy Requirements, OECD Europe, Selected Years**

| | *Production* | | *Requirements*[a] | |
|---|---|---|---|---|
| *Year* | *MTOE*[b] | *Index* | *MTOE*[b] | *Index* |
| 1957 | 375 | 100 | 557 | 100 |
| 1960 | 334 | 89 | 622 | 112 |
| 1965 | 318 | 85 | 795 | 143 |
| 1970 | 250 | 67 | 1,032 | 185 |
| 1973 | 222 | 59 | 1,178 | 211 |
| 1974 | 214 | 57 | 1,164 | 209 |
| 1975 | 221 | 59 | 1,117 | 200 |
| 1976 | 218 | 58 | 1,182 | 212 |
| 1977 | 213 | 57 | 1,196 | 225 |
| 1978 | 212 | 57 | 1,227 | 220 |

a. Excludes consumption for overseas marine bunkers.
b. Millions of metric tons of oil equivalent.

Sources: 1957—OECD, *Energy Policy: Problems and Objectives* (OECD, Paris, 1966), p. 31; 1960–73—OECD, *Energy Balances of OECD Countries, 1960/1974* (OECD, Paris, 1976), pp. 41, 46, 51, 54; 1974–78—IEA, *Energy Balances of OECD Countries, 1974/1978* (IEA, Paris, 1980), pp. 7–11.

**Table 8.9: Solid Fuel Consumption by Sector, OECD Europe, Selected Years (MTOE)**

| | *Industry* | | | |
|---|---|---|---|---|
| *Year* | *Iron & Steel* | *Other* | *Electricity Generation* | *Other*[a] |
| 1960 | 56 | 59 | 88 | 96 |
| 1965 | 55 | 47 | 110 | 92 |
| 1970 | 56 | 34 | 117 | 66 |
| 1973 | 54 | 25 | 120 | 41 |
| 1974 | 54 | 24 | 113 | 48 |
| 1975 | 44 | 21 | 112 | 41 |
| 1976 | 45 | 20 | 130 | 38 |
| 1977 | 41 | 19 | 133 | 38 |
| 1978 | 41 | 20 | 137 | 37 |

a. Mainly home heating but also includes heating of office buildings, street lighting, and agriculture.

Sources: 1960–73—OECD, *Energy Balances of OECD Countries, 1960/1974* (OECD, Paris, 1976), pp. 41, 46, 51, 54; 1974–78—IEA, *Energy Balances of OECD Countries, 1974/1978* (IEA, Paris, 1980), pp. 7–11.

coal consumption in iron and steel fell by only 2 mtoe between 1960 and 1973, it declined by 34 mtoe in all other industries. A steeper decline, 55 mtoe, occurred in the "other" sector, which consisted mainly of households and office buildings.

**Table 8.10: Structure of Fossil Fuel Consumption in Electricity Generation, OECD Europe, Selected Years (Per Cent)**

| *Year* | *Solid Fuels* | *Liquid Fuels* | *Natural Gas* |
|---|---|---|---|
| 1960 | 86 | 12 | 2 |
| 1965 | 78 | 20 | 2 |
| 1970 | 61 | 33 | 6 |
| 1973 | 52 | 37 | 11 |
| 1974 | 50 | 37 | 13 |
| 1975 | 52 | 33 | 15 |
| 1976 | 53 | 33 | 14 |
| 1977 | 56 | 30 | 14 |
| 1978 | 57 | 31 | 12 |

Sources: 1960–73—OECD, *Energy Balances of OECD Countries, 1960/1974*, (OECD, Paris, 1976), pp. 41, 46, 51, 54; 1974–78—IEA, *Energy Balances of OECD Countries, 1974/1978* (IEA, Paris, 1980), pp. 7–11.

Only in the generation of electricity did the consumption of coal rise, and this was because many electrical power stations were publicly owned. In those countries where coal production was important, these power stations, as a matter of public policy, increased their purchases of coal to compensate the industry for its loss of markets in the rest of the economy.[49] Nevertheless, coal's share of the growing electricity-generation market fell from almost 86 per cent in 1960 to slightly more than 50 per cent in 1973. The share of oil rose from about one-tenth to more than one-third. Natural gas's share also increased significantly, as is shown in Table 8.10. The newer fossil-fueled power plants became increasingly dependent on fuels other than coal.

The shares of the non-fossil fuel energy sources, hydro-power and nuclear fuel, also declined during this period, reinforcing the dependence on imported fuels. In 1960, as can be seen in Table 8.11, these non-fossil fuel energy sources accounted for more than one-half of total electric power generation. By 1973, their share had fallen to slightly more than one-third. Given the limited number of exploitable hydro sites in Europe, their share would have fallen even further had it not been for the expansion of nuclear power. However, nuclear power's contribution has been significant only since 1973, rising to one-tenth in 1978. Its rise just about kept even with the decline in hydro. Whether nuclear power would have (or, perhaps, should have) risen at a faster rate had European governments concentrated more resources on it is difficult to say. It does, however, seem fairly certain the Robinson Commission's lack of

**Table 8.11: Structure of Fossil and Non-Fossil Fuel Energy Sources in the Generation of Electricity, OECD Europe, Selected Years (Per Cent)**

| | Non-Fossil Fuels | | | |
|---|---|---|---|---|
| *Year* | *Total* | *(Hydro)* | *(Nuclear)* | *Fossil Fuels* |
| 1960 | 52 | (52) | (--) | 48 |
| 1965 | 46 | (45) | ( 1) | 54 |
| 1970 | 38 | (36) | ( 2) | 62 |
| 1973 | 35 | (31) | ( 4) | 65 |
| 1974 | 32 | (26) | ( 6) | 68 |
| 1975 | 35 | (27) | ( 8) | 65 |
| 1976 | 31 | (23) | ( 8) | 69 |
| 1977 | 37 | (27) | (10) | 63 |
| 1978 | 36 | (26) | (10) | 64 |

Sources: 1960–73—OECD, *Energy Balances of OECD Countries, 1960/1974*, (OECD, Paris, 1976), pp. 41, 46, 51, 54; 1975–78—IEA, *Energy Balances of OECD Countries, 1974/1978* (IEA, Paris, 1980), pp. 7–11.

enthusiasm for nuclear energy, together with the ready availability of cheap oil imports, did not generate strong support for its rapid development. In any event, in electricity generation, as in the other sectors of the economy, the trend towards cheap, imported energy continued unabated during the 1960s and the early part of the 1970s. That most of these imports came from the politically unstable Middle East apparently did not worry the key policy-makers of the Organisation for Economic Cooperation and Development (OECD), the successor body to the OEEC.

*OECD Study*

In a 1966 study, the OECD emphasized the need to obtain "adequate and reasonably secure supplies" of energy at "reasonable energy prices."[50] It further urged governments to adopt policies encouraging the "reduction of the cost of energy to the community as a whole."[51] Though admitting that energy imports "clearly involve direct foreign exchange costs," it nevertheless asserted that "it is not always possible to determine their net effects on the balance of payments, particularly in cases where they replace more expensive supplies from indigenous resources."[52]

The study went on to say:

> Thus, import restrictions on energy may lead to an increase in the price of energy within the country concerned which may prejudice the competitive position of its manufacturing industries, in particular of those in which energy costs are a relatively high proportion of total costs, such as the iron and steel industry. Therefore, the long-term solution to balance of payments problems would seem to lie instead either in making the indigenous industry competitive or *encouraging the transfer of labour and capital resources to more productive sectors, in particular manufactures for export*.[53] (Italics added)

Implicit in this statement is the concept that Europe reinforce its position as an importer of raw material and an exporter of manufactured products. It was, in effect, a continuation of the old colonial doctrine, and it dictated that the role of the underdeveloped countries as suppliers of cheap primary products remain unchanged. In the post–World War II world, where open imperialist transgressions were no longer socially acceptable, the method of obtaining cheap energy was to rely on the "free market" mechanism rather than on military seizure of the regions supplying raw materials. The OECD advocated a policy of economic liberalism in energy. Private enterprise was to play the key role in supplying energy and assuring the proper diversification of energy sources. "In those countries of the area," the report stated, "where oil supplies are mostly assured by private companies, market considerations have been mainly responsible for geographical diversification which has essentially been handled by private industry."[54]

Private companies, motivated by "market considerations" would presumably bring about the desirable results.

> In view of the large resources of energy available in the world as a whole, it would appear at first sight that most OECD countries could secure sufficient supplies at reasonable and stable prices by removing obstacles to international commerce and allowing each fuel to win its markets on the advantages it offered. Indeed, competition between the different fuels and freedom of choice for the consumer are among the basic elements in the energy policies of countries with or without indigenous resources.[55]

These "rules of the game," which were adopted in varying degrees by all the European OECD countries, were designed to favour oil. The thrust towards private enterprise was directed against the coal industry, which was nationalized in most European countries. The stress on "removing obstacles to international commerce," "competition between the different fuels and freedom of choice" could only benefit the international oil companies, who, because of their joint-product, were able to manipulate the prices of the individual products to achieve their competitive goals.

The OECD, of course, realized that there was a risk inherent in relying too heavily on imported energy. Pointing to the Suez crisis of 1956–57 as an example, it advocated certain safeguards against supply interruptions. One was the geographic diversification of supplies, which, as has been already mentioned, was to be accomplished by relying on the "market considerations." Another was the stockpiling of reserves, "especially of oil" to ensure "a measure of security against a temporary interruption of oil."[56] A third was the use of "reserve capacity" within the United States. This buffer was to come from the U.S. Naval Reserve, which contained "at least 165 million tons of proved reserves which are available at short notice," and from the individual states where, because of prorationing, there was "a considerable reserve capacity of oil and gas production."[57] The authors of this OECD study were seemingly unaware of the basic contradiction in their position. On the one hand, they were arguing for "competition" and "freedom of choice" for Europe while, on the other, they were looking to the United States, a country whose laws limited "competition" and "freedom of choice" in energy, to guarantee Europe's security of supply.

The use of U.S. reserves as a buffer fitted in with Bullitt's old concept to keep U.S. oil in the ground for use "as a strategic reserve." This use would not only safeguard the United States and its allies against the consequences of a prolonged interruption of oil supplies from other areas, but could also act as a powerful bargaining tool against the demands of the oil-producing countries. Any attempt by them to cut back production to receive better terms could be countered by production from U.S. reserves. The use of these U.S.-owned reserves would, at the same time, enhance U.S. control over oil-consuming areas.

The U.S. government and the oil companies were apparently able to convince the OECD that the U.S. back-up was sufficient to meet any contingencies. The 1966 OECD study therefore confidently concluded that "far from revealing signs of an imminent shortage of energy, our analysis of supply and demand suggests that ample supplies will be available for OECD countries at reasonable costs to support economic growth up to 1980 and beyond."[58]

Given these optimistic assurances, Europe began to rely more and more on imported oil. This reliance was reinforced by the automobilization of the Continent. By the late 1960s, private ownership of passenger cars and trucks had become widespread in most countries of Western Europe. Transport thus followed the other sectors in the conversion to oil. Between 1960 and 1973 consumption of gasoline increased by 200 per cent and that of gas/diesel oil by 330 per cent.[59] This shift to oil continued unabated until 1973 when, as Table 8.1 showed, oil accounted for two-thirds of Europe's fossil fuel supply.

Because of this shift, the import content of European energy requirements rose from 47 per cent in 1960 to 83 per cent in 1973. As can be seen in Table 8.12, the import content of oil, which was continuously

**Table 8.12: Share of Imports in Total Energy, Oil, and Coal Requirements, OECD Europe, Selected Years (Per Cent)**

| *Year* | *Total* | *Oil* | *Coal* |
|---|---|---|---|
| 1960 | 47 | 93 | 14 |
| 1965 | 63 | 94 | 17 |
| 1970 | 79 | 97 | 21 |
| 1973 | 83 | 99 | 21 |
| 1974 | 81 | 100 | 24 |
| 1975 | 76 | 97 | 25 |
| 1976 | 78 | 97 | 23 |
| 1977 | 76 | 94 | 24 |
| 1978 | 75 | 93 | 24 |

Sources: 1960–73—OECD, *Energy Balances of OECD Countries, 1960/1974* (OECD, Paris, 1976), pp. 41, 46, 54; 1974–78—IEA, *Energy Balances of OECD Countries, 1974/1978* (IEA, Paris, 1980), pp. 7-11.

above 90 per cent during all of the 1960s and most of the 1970s, rose from 93 per cent in 1960 to practically 100 per cent in 1974. In that year, nearly all the imports, 96 per cent, came from two geographic areas: the Middle East and North Africa.[60] Instead of the market mechanism achieving diversification in fuels and in supply sources, it achieved the opposite—the concentration on one fuel and in two geographic areas which have strong political and cultural links with each other.

### End of Oil Surplus

Along with this shift to oil came another, and somewhat unexpected, event—the disappearance of the world oil surplus. Even more ominous was the evaporation of the United States surplus. Ironically, this evaporation was partly due to the imposition of import controls, ostensibly designed to strengthen national security. In reality, they had the opposite effect, for they depleted U.S. reserves at a faster rate than would have happened if imports had been free to enter. Furthermore, they sharply accelerated the pace at which Europe converted to oil. Because of the limited access to the U.S. market, all the oil companies, internationals and independents, looked to Europe as an alternative market for their oil.

The conversion to oil lowered the elasticity of demand for that product. With the abandonment of coal-burning equipment, consumers of oil could not readily shift to other fuel sources if oil's price rose. Accompanying this decline in the elasticity of demand was a fall in the elasticity of supply. The closing down of the coal mines foreclosed competition from competing fuels. The exhaustion of reserves in other areas, especially in the United States, prevented competitors from increasing their oil production in response to an increase in price. With this lowering of demand and supply elasticities, the time was ripe for a significant increase in oil prices. The only question was when and how this increase would be instituted.

### Oil Price Increase

The Arab-Israeli war of 1973 provided the answer. If that war had not taken place, some other event would have triggered the increase, which was desired both by the international oil companies and the producing countries. In this sense their interests converged. But beyond this mutual desire for an increase, this convergence ended.

The significance of the increases of 1973–74 was that the oil-producing countries initiated them rather than the oil companies. Until the early 1970s the oil companies unilaterally determined both the price of oil and the income paid to the host countries. After 1973 the tables were turned. The oil countries determined both the price and the income accruing to the companies. Though the pricing decisions by OPEC brought significant short-term financial gains to the oil companies—their profits reached all-time highs—they also presented the companies with serious long-term problems. Essentially the assumption of unilateral decision-making by OPEC signalled a shift in the balance of power from the companies to OPEC. It represented a loss of control, and this posed a serious threat to the long-term position of both the internationals and the U.S. government.

Ironically, it was the policy, so vigorously pursued by both these groups, of converting Europe from coal to oil that gave OPEC the power it enjoys today. Had this conversion, designed to ensure U.S. hegemony over Europe, not taken place, OPEC could never have been able to assume control of oil. The action of OPEC came, furthermore, at a time when it was impossible for the U.S. to act militarily against the oil-producing countries. Bogged down in Vietnam, it was in no position to move its forces into either the Middle East or North Africa. Because of its policies, the control of oil slipped out of U.S. hands. As Robert Burns once said, "The best laid schemes o' mice and men gang aft a-gley."

Granted that Europe's shift to oil was policy-induced, one can still legitimately ask whether it was not also a necessary shift. Could the European economy have expanded as rapidly as it did, if it were mainly fueled by indigenous coal? Here the answer is not clear. There were certainly many hindrances to an increase in coal production, as the Robinson and 1966 OECD reports were quick to point out. Nevertheless, there is evidence to indicate that coal could have played a much more important role than it did. The Hartley Report, in particular, emphasized the potential for coal. Odell has pointed out that in the early 1950s, the coal mines had overcome their earlier difficulties and that both production and productivity were rising. This rise was stopped in the 1950s by the inroads of oil.[61] If, perhaps, coal had been given sufficient protection, it would have been modernized and been able to compete more effectively against oil. Coal, however, was never given that chance. It was deliberately downgraded and Europe became dependent on imported oil.

It is this dependence that is at the root of the world's present-day energy crisis. It is a crisis of control. This crisis has altered world power relationships. Above all, it has threatened the dominant position of the United States in the world of oil.

## Notes

1. C. Solberg, *Oil Power* (Mason/Charter, New York, 1976), pp. 176–77.
2. Ibid., p. 177.
3. The solid fuels include coal, lignite, and peat.
4. The fossil fuels are the hydrocarbons, *i.e.*, solid fuels, oil, and natural gas. They include neither hydroelectricity nor nuclear power.
5. United Nations Economic Commission for Europe, *The Price of Oil in Western Europe* (UNECE, Geneva, 1955), p. 28.
6. Odell, *Oil and World Power, Background to the Oil Crisis*, 3rd ed. (Penguin Books, Harmondsworth, 1974), p. 100.
7. L. Armand, *Some Aspects of the European Energy Problem, Suggestions for Collective Action* (Organisation for European Economic Cooperation, Paris, 1955), p. 35.
8. Ibid.
9. Ibid., p. 33.
10. Ibid., p. 32.
11. Ibid.
12. Ibid.
13. Ibid., p. 44.
14. Ibid., p. 36.
15. Ibid., pp. 35–36.
16. Ibid., pp. 42–43.
17. Ibid., pp. 59–61.
18. H. Hartley, *Europe's Growing Needs for Energy—How Can They Be Met?* (Organisation for European Economic Cooperation, Commission on Energy, Paris, 1956), p. 25.
19. Ibid., p. 31.
20. Ibid., p. 26.
21. Ibid.
22. Ibid., p. 73.
23. Calculated from: *Energy Balances of* OECD *Countries, 1974/1978* (International Energy Agency and Organisation for Economic Cooperation and Development, Paris, 1980), p. 8.
24. Hartley, *Europe's Growing Needs for Energy*, p. 31.
25. Ibid., p. 25.
26. Ibid., p. 27.
27. Ibid., p. 30.
28. Ibid., p. 56.
29. Armand, *Some Aspects of the European Energy Problem*, p. 46.
30. A. Robinson, *Towards a New Energy Pattern in Europe* (Organisation for European Economic Cooperation, Energy Advisory Commission, Paris, 1960), p. 5.
31. In addition to Professor Robinson, the original members of the commission were: Dr. Hermann Abs, vice-chairman; Professor Arnaldo Angeline; Walker Cisler; Dr. Henning Fransen; Professor Pierre Uri; Nicholas Vlactiopoulos; Louis Armand; and Professor Paul de Groote. The latter two were subsequently appointed as chairman and member, respectively, of

Euratom (European Atomic Energy Commission). Following their appointment, they resigned from the Energy Advisory Commission and were replaced by Jacques Desrousseaux and Dr. Anthony Trepp. See ibid., pp. 5, 9–10.
32. Ibid., p. 34.
33. Ibid., p. 60.
34. Ibid., p. 62.
35. Ibid.
35. Ibid.
36. Ibid., p. 61
37. Ibid.
38. Ibid.
39. Ibid., pp. 83–84.
40. Ibid., p. 85.
41. Ibid., p. 84.
42. Ibid., p. 85.
43. Ibid., p. 60.
44. Ibid., p. 28.
45. Ibid., p. 5.
46. Ibid., p. 5.
47. W. Jensen, *Energy and the Economy of Nations* (G.T. Foulis, Henley-on-Thames, 1970), pp. 108–9.
48. *Energy Policy, Problems and Objectives* (Organisation for Economic Cooperation and Development, Paris, 1966), p. 31; *Energy Balances of* OECD *Countries 1960/1974* (Organisation for Economic Cooperation and Development, Paris, 1976), pp. 41–55; *Energy Balances of* OECD *Countries, 1974/1978*, pp. 7–10.
49. Jensen, *Energy and the Economy of Nations*, p. 137.
50. *Energy Policy, Problems and Objectives*, p. 104.
51. Ibid.
52. Ibid.
53. Ibid., p. 122.
54. Ibid., p. 111.
55. Ibid., p. 104.
56. Ibid., p. 111.
57. Ibid., p. 112.
58. Ibid., p. 143.
59. Calculated from: *1977 Oil Statistics* (IEA, Paris, 1977), Table D.
60. Calculated from: Ibid., Table A.
61. Odell, *Oil and World Power*, p. 104.

# CHAPTER NINE

# The Erosion of U.S. Dominance

U.S. dominance, if we recall, reached its peak with the restoration of the Shah in Iran in 1954. Not only did U.S. corporations control most of the Middle East oil but they also commanded key positions in the economies of most of the advanced capitalist countries.

## Reasons for Erosion

This U.S. dominance was, however, relatively short-lived. It began to erode during the late 1950s and the early 1960s. There were many reasons for this gradual erosion.

### *Socialist Countries*

One was the growing economic strength of the Socialist countries. This strength enabled them both to resist U.S. attempts to control their economies and to offer Third World countries an alternative to the U.S. multinationals. Soviet assistance to Cuba, for example, provided that country with the means to withstand the U.S. economic blockade.

### *European and Japanese Revival*

A second reason was the revival of the war-shattered economies of Western Europe and Japan. Having recovered from the effects of the war by the mid-1950s, they began to compete fiercely with U.S. companies for world markets. Their expansion, interestingly enough, was partially triggered by the U.S. policy of providing cheap oil to these countries as a means of displacing coal. This cheap energy gave European and Japanese firms a competitive advantage over U.S. firms, who, because of oil import quotas, had to pay higher prices for their energy inputs.

The economies of Japan and west European countries grew at a more rapid rate than the United States economy. As can be seen in Table 9.1, the U.S. growth rate between 1960, when its major rivals began making a significant impact on the world economy, and 1973, the year of the oil

**Table 9.1: Average Annual Growth Rates in Real Gross National Product, 1960–73, Major Industrialized Market Economies (Per Cent)**

| *Country* | *Rate* |
|---|---|
| Japan | 10.5 |
| France | 5.7 |
| Canada | 5.4 |
| Italy | 5.2 |
| Germany | 4.8 |
| United States | 4.2 |
| United Kingdom | 3.2 |

Source: *Economic Report of the President Transmitted to the Congress January 1981* (U.S. Government Printing Office, Washington, 1981), p. 353.

crisis, was lower than that of any other important industrialized market economy with the exception of the United Kingdom. Japan's rate, for instance, was approximately $2^1/_2$ times greater than the U.S. one.

### DECLINE IN U.S. EXPORT SHARES

Exports played a major part in this growth surge. European and Japanese products began to capture the lion's share of the increasing volume of world trade. The U.S. share of world manufactured exports fell from 25 per cent in 1955 to 16 per cent in 1973.[1] Given the extensive U.S. investments in overseas manufacturing facilities, this decline is, in one sense, not surprising. To the extent that U.S. firms began producing abroad items which they had previously exported from the States, the U.S. export share was bound to fall. This decline thus did not necessarily signify a deterioration in the U.S. competitive position. During this period, however, non-U.S. multinationals grew at a faster pace than U.S. firms.[2] This differential growth rate was a significant factor in lowering the U.S. share. Following their gains in foreign markets, the Europeans and Japanese then invaded the U.S. market. As a result the U.S. surplus on merchandise trade fell precipitously. As was shown in Table 5.5 (page 84) the surplus turned into a deficit in the 1970s. This decline, as already mentioned, took place before the rise of oil prices and was due mainly to an increase in manufactured imports.

### BALANCE OF PAYMENTS DEFICIT

This fall in the trade balance further aggravated the U.S. balance of payments deficit, which began appearing in the late 1950s. The country's large trade surplus in that period was insufficient to finance both

U.S. military expenditures and investments abroad. The U.S. paid for this deficit with its gold and foreign currency assets. The drain on its international reserves became so pronounced that in 1971 President Nixon broke the link between gold and the U.S. dollar. This link, which was the cornerstone of the postwar international monetary system, set the value of gold at $35 an ounce. The United States pledged to sell its gold to all governments and central banks of countries who were members of the IMF at that rate. Under this arrangement, the dollar was considered "as good as gold."

The large and persistent U.S. balance of payments deficits brought an ever-increasing flow of dollars into the treasuries and central banks of foreign governments. As this flow continued, many governments began to lose confidence in the dollar. Accordingly, they traded their dollars in for gold. The U.S. gold stock consequently fell from a high of $25 billion in 1949 to $10 billion in 1971,[3] the year the U.S. broke its pledge to exchange gold for dollars at $35 an ounce. This action led to a restructuring of the international monetary system from one based on fixed exchange rates to a system of flexible rates.

Another important outcome of the U.S. balance of payments deficits was a major redistribution of international reserves, which consisted of Special Drawing Rights (SDRs), positions in the IMF, and holdings of "hard" currencies as well as gold. As Table 9.2 indicates, the U.S. controlled more than half of the world's reserves in 1949. By 1971, the U.S. share had fallen to 10 per cent. In 1973, it had declined to 8 per cent.

**Table 9.2: U.S. Share of International Reserves, Selected Years**

| *Year* | *Total Reserves* ($ Billions) | *U.S. Reserves* ($ Billions) | *U.S. Share of Total* (%) |
|---|---|---|---|
| 1949 | 46.1 | 26.0 | 56 |
| 1952 | 49.3 | 24.7 | 50 |
| 1962 | 62.6 | 17.2 | 28 |
| 1968 | 77.4 | 15.7 | 20 |
| 1969 | 78.3 | 17.0 | 22 |
| 1970 | 92.6 | 14.5 | 16 |
| 1971 | 130.6 | 13.2 | 10 |
| 1972 | 158.7 | 13.2 | 8 |
| 1973 | 183.7 | 14.4 | 8 |

Source: International Monetary Fund, *International Financial Statistics*, various issues.

The beneficiaries of this redistribution were America's major industrial competitors. By 1971, the reserves of both Germany and Japan exceeded those of the United States. Germany's were \$19 billion, Japan's were \$15 billion, while those of the United States were \$13 billion. By that year more than half the world's reserves were concentrated in the countries of Western Europe.[4] It was obvious that the United States no longer held sway over the non-Communist industrialized world.

*Cold War*

The Cold War played an important part in this process. Ironically, U.S. strategy in the Cold War was not only to contain the Soviet Union but also to enforce U.S. hegemony over the rest of the world. But the Cold War brought the opposite results.

Among other things, the Cold War adversely affected the U.S. export position. Not only did it generate inflationary pressures in the United States, thereby undermining the competitiveness of U.S. exports, but it also slowed down the rate of technological innovations in market-oriented goods. As Solo has noted, most of the expenditures in the United States on research and development, the key to technological advancement, were on military projects. He pointed out that

> the increase in space and military R & D may be a diversion from and, consequently, may be at the expense of R & D oriented toward economic growth. And although the ratio of R & D expenditures to national income has increased greatly since 1953, the corresponding trend of growth-oriented R & D seems actually to have been downward. . . . Not only does military and space R & D divert research power from its growth-oriented counterpart, but R & D per se can draw creative resources from alternative forms of inventive and innovative activity. . . . Through the advent of R & D, those who were, or who might have been, restless, probing industrialists, innovating entrepreneurs, or inventors tinkering in their shops became, instead, engineers on project teams, heads of research divisions, scientists in laboratories, or subcontractors with the task of developing a component for a complex weapons system. Something is given to one side, but lost from the other.[5]

It is interesting to note that the two countries that turned in the best export performances during the 1960s and early 1970s, Germany and Japan, were those whose military expenditures were limited by prohibitions imposed on them after the war. Neither country has allocated nearly the proportion of its resources to the military as the United States and some of its principal allies have.

The losses in exports were compounded by the extensive expenditures to maintain facilities and bases overseas. These expenditures, as well as those incurred in the Vietnam war, transferred international reserves from the United States to other countries, thereby further eroding America's international economic power base.

## Implications of Erosion

Until the early 1970s the U.S. held one important trump card against this erosion—the control of oil. With the world becoming increasingly dependent on oil, the U.S. was in a position to keep recalcitrant rivals in line. The loss of control in 1973 undermined this pillar of the American Empire. This was the essence of the oil crisis as far as the U.S. was concerned. The problem was succinctly stated by Walter Levy in the following words:

> In the past, control over the international oil companies could be and was used as a political instrument by their home countries in their relationship to importing countries, such as the United States apparently did during the first Suez crisis or say for oil trade with Cuba, and so on. This possibility, to use the control over foreign oil for political-strategic purposes, is disappearing fast.[6]

### *The New Entrants*

What the U.S. was facing was the rise of new entrants onto the world oil scene. They were not the Schumpeterian innovators who deluged the market with new products. They were in fact imitators, who took over a product introduced by others. Nevertheless, they were entrepreneurs, in the true sense of the word, modern-day Rockefellers. They had recognized a change in world market conditions, seized the opportunities presented by that change, and elevated themselves to positions of world power. As Levy put it, the world was witnessing "a shift of the major center of power over international oil from the home countries to the producing countries."[7]

The implications of this shift were enormous. Not only did it give the OPEC countries control over a physical resource, oil, but it also gave them control over immense financial resources. Since the price rises of 1973–74, an increasingly greater proportion of international reserves has been accruing to the oil-producing countries. As Table 9.3 shows, the portion of international reserves accruing to the oil-exporting countries increased spectacularly during the 1970s, rising from 6 per cent at the end of 1970 to more than 20 per cent at the end of 1980. The oil-

**Table 9.3: Oil-Exporting Countries' Shares of International Reserves, Selected Years (Per Cent)**

| *Year* | *Oil Countries* | *Saudi Arabia* |
|---|---|---|
| 1970 | 6 | 1 |
| 1973 | 8 | 2 |
| 1978 | 17 | 5 |
| 1980 | 21 | 6 |

Source: International Monetary Fund, *International Financial Statistics*, various issues.

exporting countries have gained control of approximately one-fifth of the non-Communist world's international reserves. Saudi Arabia, alone, possessed the same proportion of reserves at the end of 1980 that all the oil-exporting countries had had at the end of 1970. Its reserves were about equal to those of such industrialized countries as the United States, Japan, and Italy, and exceeded those of Great Britain.

This acquisition of reserves has meant that a new bourgeoisie, a bourgeoisie whose wealth is derived from mineral resources rather than from industrial production, has entered the arena of world capitalism. As such, it poses a threat to the established wealth. As this bourgeoisie accumulates greater reserves, it increases its ability to buy out the established bourgeoisie and to compete actively against them by setting up enterprises to rival theirs.

### *The Threat to U.S. Investment*

In countries which, because of sparse populations or fear of social change, are unable to absorb large injections of developmental expenditures, there is little choice but to invest these reserves in the developed capitalist economies. Carried to the extreme, these investments could be large enough to give the oil-exporting countries control over the advanced capitalist economies. In a memorandum prepared for the 1974 pre-summit meeting of economists for the White House Conference on Inflation, Levy outlined the implications of this development.

> In the year ending June 30, 1975, conservative estimates imply that OPEC countries may add between $60 and $75 billion to their surplus holdings of foreign currencies. Amounts of that order would be equal to about 75 per cent of the total book value of U.S. direct foreign investments, accumulated over many decades and owned by many hundreds of private corporations, whereas this OPEC surplus, held by a handful of governments, will arise in

only these current twelve months. The annual amount substantially exceeds total U.S. net foreign public and private long-term and short-term liquid and non-liquid assets. By July, 1975 gold and foreign exchange reserves of a handful of OPEC countries will probably be larger than those of all major industrialized nations combined. From then on—until such time, if ever, as these oil-exporting economies can absorb all their oil revenues—their surplus claims on the rest of the world will go on pyramiding.[8]

Though Levy erred in his estimates of the magnitudes—the OPEC countries did not accumulate the massive reserves he had predicted—he nevertheless was correct about the trends. Surplus claims would continue to pyramid, though at a slower rate than Levy thought. This pyramiding could eventually place the OPEC countries in a position to control the economies of the non-Communist countries. Levy therefore argued that the industrialized countries must develop a coordinated strategy to counter this threat.[9]

Levy was, of course, speaking from a United States point of view. He recognized that the United States now faced a third threat to its hegemony. The first was that posed by the Communist countries. The second was the one flowing from the increasing economic strength of Europe and Japan. And the third was the rise in the power of the oil-exporting countries. Probably the greatest fear of Levy and the other influential U.S. policy-makers was that these diverse groups might find it to their interests to join forces against the United States.

## U.S. Reaction to Erosion

To prevent this realignment from happening, U.S. strategists were determined to forge a united front of the non-Communist industrialized countries against OPEC. As early as 1969 the U.S., anticipating the problems that might arise from the decline in the oil surplus under its control, tried to persuade the members of the OECD oil committee to formulate a common approach to energy problems. Their response was somewhat less than enthusiastic, although the U.S. suggestion was supported by England and Holland, the home countries of British Petroleum and Shell.[10] In subsequent years the United States made several other attempts to achieve a coordinated policy but with no success. In early 1973, before the advent of the oil crisis, James Akins, then director of the Office of Fuels and Energy of the Department of State, wrote a provocative article in *Foreign Affairs* in which he stated the U.S. arguments for such a policy. He advocated "the formation of an international authority" which would "avoid cutthroat competition for available energy in terms of shortage." He feared that "such competition could drive prices far higher than we can presently imagine."[11]

### *Consumers' Cartel*

He therefore advocated the establishment of a consumers' "cartel," as a form of countervailing power to OPEC. One of the tasks of this "cartel" would be to support the international oil companies in their negotiations with OPEC. "The producer governments," he stated, "have banded together in a well functioning organization. Their immediate adversaries are only the companies—an unequal contest."[12]

This alleged impotence of the oil companies was, of course, a hyperbole, similar to Adelman's assertion that the companies had become mere tax collectors for OPEC.[13] These companies have for years been powers in their own right. Furthermore, they have always had the backing of their home governments in foreign ventures. Levy himself has admitted they had benefited for decades "from the immense power of the United States."[14] This backing was reinforced early in 1971 when the State Department persuaded the U.S. Justice Department to exempt the oil companies from antitrust prosecutions if they collaborated with each other in their dealings with OPEC.[15]

Equally dubious was Akins's implication that the oil companies could act in the interests of consumers and that they would not raise prices on their own. "Many consumers already believe," he wrote, "that the companies have not been adequately vigorous in resisting producer demands, as they could and usually did pass on to the consumer any tax increases."[16] He therefore urged that, in the interests of "consumer solidarity," all the governments of the major oil-importing countries, not merely the home governments of the international oil companies, give backing to these companies to enable them to resist OPEC's demands. Such backing would, of course, bolster the position of these companies and their home governments, especially the government of the United States. The last thing that United States policy-makers wanted was for the importing countries to bypass the oil companies and deal directly with the governments of the producing countries.

After the price escalation of 1973–74, the danger that individual countries would make separate deals with the oil countries became more pronounced. It was for this reason that the United States opposed the French suggestion for a meeting between the consuming and producing nations. It insisted instead that there must be an agreement among the consuming countries before any such dialogue took place. Secretary of State Henry Kissinger stated quite categorically that "we will not go to a consumer-producer conference without prior agreement on consumer cooperation."[17] Without such a "prior agreement," separate deals could be negotiated between individual consumer and producer countries. After much wrangling the French accepted the concept of a "prior agreement" at a meeting in Martinique between Presidents Ford and Giscard d'Estaing at the end of 1975.[18]

### *International Energy Agency*

The French, however, refused to go along with the U.S. suggestion to form a consumers' "cartel." This "cartel," the International Energy Agency (IEA), was formed in 1974 by the major consuming nations, France excepted, for the purpose of sharing oil supplies in the event of a cut-off directed against either the group as a whole or any of its members.[19] Not only was it designed to prevent consumers from bidding up the price in times of shortages, but it was also designed to protect individual countries against a selective oil boycott. The 1973–74 boycott, it must be remembered, was directed primarily against the United States and Holland in retaliation for their support of Israel. The IEA agreement therefore gave an implicit carte blanche backing to U.S. Middle East policies by assuring the U.S. that it would not have to bear the full burden of any adverse reaction to its policies.

#### OIL SHARING

Article 17 of the agreement, for instance, invokes the oil sharing provisions if

> any Participating Country sustains or can reasonably be expected to sustain a reduction in the daily rate of its oil supplies which results in a reduction of the daily rate of its final consumption by an amount exceeding 7 per cent of the average daily rate of its final consumption during the base period . . .[20]

Once Article 17 is invoked, all the members of the IEA are required to allocate to the affected country a quantity of petroleum and products "equal to the reduction in its oil supplies which results in a reduction of final consumption over and above" 7 per cent.[21] Furthermore, this compulsory allocation cannot be effectively vetoed if the Secretariat and the Management Committee decide that a country is eligible to receive it. It comes into force automatically unless "the Governing Board, acting by special majority, . . . decides not to activate the emergency measures."[22] This "special majority" requires the consent of all but two members.[23] Thus, if as few as three of the original sixteen members vote to sustain the recommendations of the Secretariat and the Management Committee, the allocation provisions come into immediate effect. This provision effectively protects any country, and the United States in particular, against a selective boycott. If the allocation procedures work according to plan, any individual country would suffer no more than a 7 per cent fall in consumption. This provision removed an important constraint on the United States in its dealings with the OPEC countries, by guaranteeing the country an assured source of supply in the event of a boycott. The *Oil and Gas Journal* commented thus:

> In theory, U.S. domestic oil production is available for international allocation under the emergency program, the State Department notes. In practice, however, a spokesman says, only under the most extreme emergency situation would the U.S. ever be called upon to share any of its domestic production with the other IEA countries.[24]

The journal then quoted Julius Katz, acting assistant secretary of state for Economic and Business Affairs, as saying that "in the event of a selective embargo against the U.S. which cut back our available oil by more than 7 per cent, we would of course expect to receive oil from the other member countries."[25]

Article 17, while protecting the U.S. against a selective boycott, actually enhances the chance of a general boycott. Given the protection afforded to the United States, the OPEC countries could retaliate against the U.S. only by embargoing oil to all countries that are members of the IEA. This general embargo would be the sole effective way of reducing significantly the oil supplies flowing to the U.S.

The fear of being forced to underwrite America's Middle East policies made several nations wary of joining the IEA. Significantly, Norway, the only European country with a surplus of oil at that time, did not sign the original agreement and joined the IEA later on an associate basis, which exempted her from the energy-sharing provisions.[26] Canada, another important producer of oil, did sign the agreement. Nevertheless there were many questions raised in Parliament about the wisdom of adhering to it. Most important of all, France, which has pursued its own Middle East policies, refused to join.

### ROLE OF THE OIL COMPANIES

Another interesting feature of the IEA is the role it assigns to the oil companies. These companies are to play an important part in carrying out any emergency allocation program. Article 6, for instance, stipulates that the Governing Board will decide "on the procedures and modalities for the participation of oil companies" in any allocation scheme.[27] Chapter VI establishes a "framework for consultation with oil companies."[28] Chapter V, while requiring oil companies to provide relevant information to their home governments, who in turn will make such information available to the IEA, also allows them to withhold confidential data and proprietary information, such as "patents, trademarks, scientific or manufacturing processes or developments, individual sales, tax returns, customer lists or geological and geophysical information, including maps."[29] These confidentiality and proprietary provisions allow both the oil companies and their home governments considerable discretion in selecting the information to be submitted to the IEA. Finally, Articles 30 and 35 stipulate that the Standing Group on the Oil Market and the Standing Group on Emergency Questions shall, in preparing their respective reports, "consult with oil companies to ensure that the System is compatible with industry operations."[30]

These provisions give the oil companies enormous power. As the *Oil and Gas Journal* pointed out, "the international oil companies will have a key role in planning and carrying out any measures to share and redirect tankers and supplies if oil imports are curtailed again . . ."[31] Their power is exercised through the Industry Advisory Board. This board, which originally consisted of representatives of fifteen oil companies, "[provides] the Agency with a range of advice on matters related to the emergency allocation procedures and emergency reserves."[32] This working group of companies is headed by Exxon,[33] the successor to Rockefeller's Standard Oil (New Jersey).[34] The board enables Exxon to coordinate activities with its supposed competitors. The antitrust implications of these moves were not lost on the U.S. government, which accordingly gave the participating companies antitrust clearance.[35] The U.S. government thus gave its official blessing to a new cartel of oil companies, which is to act as a counterweight to the cartel of oil-producing countries.

The provisions of the IEA agreement also put the oil companies in a position of forcing the other governments to back them. This backing, as was noted earlier, is one of the principal objectives of U.S. policy, to be achieved through the companies' control of the flow of information. The information on which the IEA must base its decisions originates with the oil companies, who pass it on to their home governments—essentially the United States, England, and Holland—who, in turn, give it to the

Secretariat of the IEA. Furthermore, in their consultations with the various IEA committees, the oil companies can put their own interpretation on this information flow. Since the oil companies are not disinterested observers of the IEA decision-making process, they obviously have a stake in presenting their home governments and the IEA with their own viewpoint. Their home governments, and especially the U.S. government, also have the opportunity of influencing IEA decisions by the control and interpretation of information flows. This process has serious implications for IEA decision-making. It has the potential of repeating the situation which led the European governments, after consultation with the oil companies, to abandon coal for oil.

The dominant position of the oil companies is further reinforced by the principles underlying the allocation mechanism. Article 9, for instance, reads:

> When allocation takes place, an objective of the Program shall be that available crude oil and products shall, insofar as possible, be shared within the refining and distributing industries as well as between refining and distributing companies in accordance with historical supply patterns.[36]

This provision protects the historical position of the international oil companies. No emergency would be permitted to challenge this position. At the same time, there is no effective provision to prevent these companies from using an emergency as an excuse for raising prices. Aside from a call for "basing the price for allocated oil on the price conditions prevailing for comparable transactions"[37] and an article stating that "questions relating to the price of oil allocated during an emergency shall be examined by the Standing Committee on Emergency Questions,"[38] there are no references to pricing policies. These policies will presumably be decided by the Standing Committee on Emergency Questions after consultation with the oil companies. Given these circumstances, one can only wonder how effective this Standing Committee would be in protecting consumers' interests in an oil cut-off.

### FLOOR PRICE

Though ostensibly formed to act as a consumers' "cartel," the IEA apparently was not interested in forcing oil prices down too dramatically and certainly did not wish them to return to the levels prevailing in early 1973. There were both political and economic reasons for this reluctance.

Kissinger expressed the political reasons when he warned about the effects of a sharp price decrease on two of America's most important client oil states, Saudi Arabia and Iran. In an interview held in early 1975, he expressed his fears in the following terms:

> The only chance to bring oil prices down immediately would be massive political warfare against countries like Saudi Arabia and Iran to make them risk their political stability and maybe their security if they did not cooperate. *That is too high a price to pay even for an immediate reduction in oil prices.*
>
> If you bring about an overthrow of the existing system in Saudi Arabia and a Khadaffi takes over, or if you break Iran's image of being capable of resisting outside pressures, you're going to open up political trends that could defeat your economic objectives.[39] (Italics added)

In other words, Kissinger was willing to pay a price for political stability in the Middle East. Though unquestionably concerned about the implications of the shift in world power following the rise in oil prices, he nevertheless felt that the consequences of opposing this shift would be worse. He looked upon the IEA as a means of preventing further price increases rather than as an instrument to bring about decreases.

At the end of 1974 there was a fairly widespread belief that the price of oil would fall. A number of factors contributed to this belief. Among them were (1) the steps by the IEA to decrease oil imports, (2) the recession which hit the non-Communist world following the oil price increases, (3) the growing quantities of oil owned directly by the oil countries as a result of participation and nationalization agreements, and (4) the inability of the OPEC countries to coordinate production.

Following the formation of the IEA, the participating countries agreed to reduce imports in 1975 by two million barrels per day from their 1973 levels. They also proposed further reductions from 1973 of 8 and 6 per cent in 1976 and 1977 respectively.[40] The reduction in imports exerted downward pressure on prices.

This downward pressure was intensified as a result of the fall in economic activity. In 1974, for instance, the rate of growth in real Gross National Product in the OECD countries fell to 0.2 per cent from 5.9 per cent in 1973. By 1975, it was equal to -0.9 per cent, indicating that the actual level of output had fallen.[41] As a consequence, total energy consumption fell by 0.8 per cent in 1974 and 2.8 per cent in 1975,[42] while oil imports dropped by 3.1 per cent and 6.7 per cent, respectively, in those years.[43]

These developments created a surplus of oil in the sense that more oil could be produced than sold at the prevailing price. By the end of 1974 the OPEC nations, according to Henry Kissinger, had shut down one-quarter of their capacity, approximately nine billion barrels per day, in order to eliminate this surplus.[44]

A great deal of the oil produced was owned directly by oil countries as a result of the spread of participation agreements, under which the oil countries assumed an equity share in the companies, and outright nationalizations. Many believed that the OPEC countries, in their eagerness to sell this crude, would cut prices. Levy even suggested that it might be a good idea for the oil companies to abandon their concessions and "reappear as competitive buyers of crude from producing countries." He argued that if the oil states took possession of these concessions then "one might expect that OPEC unity would erode and the producing countries would eventually compete with each other for export sales to the companies,"[45] thereby causing prices to fall.

What Levy pointed out was that the coordinating role of the international oil companies was eroding as the role of the national oil companies in the producing countries was expanding. As Blair has noted, the internationals have for years coordinated production as a means of stabilizing prices.[46] The erosion of their control brought the individual oil countries into the world market as competitors.

By early 1974, the combined effects of IEA's import cutbacks, the recession, the increasing quantities of oil in the hands of the oil countries, and the lack of coordination among these countries began to make themselves felt. In late January, Saudi Arabia served notice that it was going to seek a reduction in the oil price.[47] In March, OPEC announced that it was going to freeze prices for three months, commencing April 1.[48] In July, Kuwait rejected all bids at an auction for its oil because they were too low.[49] In September, Texaco and Standard Oil of California cut production in Indonesia by 200,000 barrels a day in response to weakened market conditions.[50]

Reviewing these developments, Frank Gardner, a columnist for the *Oil and Gas Journal*, gleefully predicted in early 1975 that oil prices would soon come tumbling down. Gardner wrote:

> OPEC's one big objective since its inception was to shore up crude prices. That has now been done—overdone. No product can long continue to sell at a hundred times its production cost, especially when it's in excess supply. The competition is coming, and with it, lower crude prices—not at the lows of 1960, but lower.[51]

The prospect of falling oil prices threatened to undermine the IEA's objectives of decreasing dependence on imports from the OPEC countries. The agreement establishing the IEA called for cooperation by the participating countries in the "development of alternative sources of energy such as domestic oil, coal, natural gas, nuclear energy, and hydroelectric power."[52] The development of these sources required investments whose profitability would be endangered by low OPEC oil prices. These investments had to be protected against this contingency. There were, therefore, strong "economic" reasons for opposing a return to low oil prices.

For these reasons Kissinger urged that consuming countries impose a floor price on imported oil. He asked that the IEA members "not allow imported oil to be sold domestically at prices which would make these new sources noncompetitive."[53] He particularly wanted the IEA to prevent OPEC from engaging in "predatory price cutting" in order to discourage investments in new energy sources.[54] His views were echoed even more strongly by Thomas Enders, then U.S. representative to the IEA and later U.S. ambassador to Canada, who told an oil conference in Colorado in December 1975 that:

> Consuming nations can act together to make it more probable that the bulk of available energy opportunities are in fact exploited, by protecting them against the downside risk of predatory pricing by OPEC or *a return to competitive conditions in the international oil market*.[55] (Italics added)

This aversion to competition stands in marked contrast to the U.S. attitude before the escalation of oil prices. As pointed out in Chapter 8, the U.S., in its eagerness to have Europe convert from coal to oil, continually stressed the desirability of freedom of choice for consumers and the benefits of competition in determining the forms of energy use. Such policies coincided then with the interests of the international oil companies. In the post-escalation era, the U.S. cautioned against the "return to competitive conditions," which position again coincided with the interests of these companies. Competition apparently was desirable when it benefited these companies and undesirable when it did not.

In any event the U.S. vigorously sought to have the other IEA members agree to a floor price. When it first broached the matter, it was rebuffed, because the other members were interested in an immediate reduction in high energy costs. Japan, a country with few indigenous energy resources, was most vehemently opposed to the concept. The IEA thus rejected the floor price at its meeting of March 1975.[56]

Nevertheless, the United States persisted in pressing for such a price. Finally, in December 1975, the IEA adopted a floor price of $7 per barrel, but not without misgivings from some of its members. There was particular objection to the relatively high floor price, which would saddle the consuming countries with heavy energy costs. As the *Oil and Gas Journal* reported:

> Some see it as an attempt to divert investment into non-OPEC energy sources. Others see it as an endorsement of OPEC action of October 1973 when prices were doubled. A $7 floor price is almost three times the price of October 1973.[57]

The floor price was part of the "Long-Term Co-operation Programme," which stipulated that "the Participating countries shall, as a general measure of co-operation, ensure that imported oil is not sold in their domestic markets below a price corresponding to U.S. $7/bbl."[58] This price, which was called the MSP (Minimum Safeguard Price), could be changed by unanimous vote of the Governing Board.[59] Presumably it would be raised if it were insufficient to assure the profitability of investments in alternative energy sources.

The IEA members were supposed to take specific measures if the price of imports fell below the MSP. These measures included the imposition of specific and variable levies, import quotas, and consumption taxes.[60] These provisions served as a disincentive for OPEC to make "legitimate" price cuts, *i.e.*, a price reduction in response to change in supply-demand relationships. If these cuts were taxed by the oil-importing countries, the OPEC countries would have little chance of increasing sales through price reductions. These reductions would merely transfer income from the OPEC to the consuming countries, and OPEC, rather than the oil companies, would bear the full brunt of any reduction in price.

### INVESTMENT INCENTIVE

By imposing a floor price, the IEA transformed itself from a consumers' "cartel" to a vehicle for protecting the interests of the energy companies, to which the consumers' interests in lower prices were made subservient. This transformation was evident in the importance attached to stimuli to investment in the Long-Term Co-operation Programme. Chapter III, for instance, states that "the Participating Countries will, as appropriate and necessary, create a climate favourable for investment in energy."[61] A "favourable climate" is, of course, one that is highly profitable. Perhaps more important than protection for investment in general is the *de facto* protection it gives to investment by American companies in other IEA countries.

That this was one of the purposes of the Long-Term Co-operation Programme was made clear by Enders, who stated:

> Each of the consuming countries faces an enormous task in mobilizing the capital required to diminish its dependence on imported energy. . . . But consuming countries can assist each other by making it possible for countries that are capital-rich but poor in energy resources to invest in large-scale projects in other consuming countries.[62]

The U.S. was "capital-rich," but "energy-poor" in the sense that it was a net importer of energy. Its companies would certainly look forward to investing in countries which do not have the abundant capital but are relatively rich in energy resources—like Canada, Norway, and the United Kingdom.

Many provisions of the Long-Term Co-operation Programme facilitate this movement of U.S. companies into other IEA countries. One provision stipulates that the host governments "accord national treatment and most-favoured-nation treatment" to foreign enterprises engaged in energy projects.[63] Another calls on the host countries to "avoid the introduction of new limitations upon the extent to which such enterprises are accorded national treatment and most-favoured-nation treatment."[64] These provisions are merely an institutionalization of the "open door" doctrine which the U.S. used in its disputes with Britain and Holland in the 1920s. This doctrine is reinforced by another provision calling upon the host governments to consider "affording to nationals and companies of other Participating Countries which become parties to such a project incentives similar to those granted to its own nationals of [sic] companies."[65] This provision constrains a country like Canada, which wants to reduce the foreign control of its economy, from taking steps to encourage its own nationals and companies in energy projects.

Canada had already taken steps to reduce the inflow of foreign investment. In 1974 it established the Foreign Investment Review Agency (FIRA) to decide "whether there is or will be significant benefit to Canada in proposals by non-Canadians regarding acquisition of control of Canadian business enterprises or the establishment of new businesses in Canada."[66] FIRA has the right to veto any foreign investment which it deems contrary to Canada's interests. As was to be expected, Canada objected to Chapter V of the Programme, which calls on IEA members to "work towards the identification and removal of legislative and administrative measures which impair the achievement of the overall objectives of the Programme."[67] Because of its objections, Canada was exempted from Chapter V. But its provisions apply to all other signatories.

This "open door" policy towards investment originating in IEA member countries stands in marked contrast to the "closed door" treatment of investments from other areas, like the OPEC or the Communist countries, to whom the non-discriminatory provisions do not apply. As in the 1920s, they are used primarily to enhance the position of U.S. companies.

Another interesting feature of the Programme is its treatment of taxes. It calls on the "host governments to take account in considering alterations of taxation and production policy as they relate to those projects of the effect of such alterations on the economics of such projects including projects already underway."[68] This is an updated version of the "minimum duty" doctrine which the United States used against Mexico in the post–World War I years. Under it, no country should apply to these particular projects a tax that it applies to the rest of the economy if this tax adversely affects the profitability of the projects. This "minimum duty" doctrine is thus one which discriminates in favour of foreigners. As in the 1920s, the U.S. uses this doctrine or its opposite—the supposedly non-discriminatory "open door" one—whenever it suits its needs.

Probably the most revealing provision of the Programme is the one relating to the exports of energy resources from the host countries. This provision requires that the host country consider

> guaranteeing to enterprises or other parties from other Participating Countries, through legislative and administrative actions if necessary and appropriate, the right to export to their own countries a portion of the product of the project corresponding to their participation in the project or an equitable portion as may be agreed.[69]

This paragraph, along with those encouraging the international flow of investments among the IEA countries, fits in with Enders's aim of enabling the "capital-rich" but energy-poor nations of the IEA to gain access to the other members' resources through investment. The U.S. is, in effect, using the IEA as a means both of halting the general erosion of its international position and of regaining control over energy resources.

## Notes

1. *National Institute Economic Review*, no. 12 (1960), p. 63, and no. 70 (1974), p. 93.
2. S. Hymer & R. Rowthorn, "Multinational Corporations and International Oligopoly: The Non-American Challenge" in C. P. Kindleberger (ed.), *The International Corporation* (The M.I.T. Press, Cambridge, Massachusetts, 1970), pp. 57–91.
3. *Economic Report of the President Transmitted to the Congress February 1974* (U.S. Government Printing Office, Washington, 1974), p. 356.
4. Ibid., p. 355.
5. Robert Solo, "Gearing Military R & D to Economic Growth" in J. L. Clayton (ed.), *The Economic Impact of The Cold War, Sources and Readings* (Harcourt, Brace & World, Inc., New York, 1970), pp. 157–64.
6. W. J. Levy, "An Atlantic-Japanese Energy Policy," *Foreign Policy*, no. 11 (1973), p. 176.
7. Ibid., p. 171.
8. W. Levy, "The Impact of Exploding Oil Costs on the World Financial and Economic System," based on a statement prepared for the pre-summit meeting of economists for the White House Conference on Inflation, September 1974, p. 4.
9. Ibid., p. 12.
10. J. E. Akins, "The Oil Crisis: This Time the Wolf is Here," *Foreign Affairs*, vol. 51, no. 3 (1973), p. 486.
11. Ibid., p. 487.
12. Ibid., p. 488.
13. M. A. Adelman, *The World Petroleum Market* (The Johns Hopkins University Press, Baltimore, 1972), p. 250.
14. Levy, "An Atlantic-Japanese Energy Policy," pp. 161–62.
15. Akins, "The Oil Crisis: This Time the Wolf is Here," p. 473.
16. Ibid., p. 488.
17. "Kissinger on Oil, Food, and Trade," *Business Week*, January 13, 1975, p. 67.
18. D. Evans, *The Politics of Energy, The Emergence of the Superstate* (Macmillan of Canada, Toronto, 1976), p. 51.
19. OECD, "Establishment of an International Energy Agency in the OECD," Press Release A (74) 48, Paris, November 15, 1974. The original members of the IEA were Austria, Belgium, Canada, Denmark, Germany, Ireland, Italy, Japan, Luxembourg, the Netherlands, Spain, Sweden, Switzerland, Turkey, the United Kingdom, and the United States. Of the OECD members, France and Norway were the significant non-members of the IEA. Norway, as a potentially large oil producer and exporter, had doubts about joining an organization representing consumers' interests. Norway later joined under special arrangements. France has never joined.
20. *Agreement on an International Energy Program* (International Energy Agency, Paris, 1975), p. 15.
21. Ibid., p. 12.

22. Ibid., p. 16.
23. Ibid., p. 34.
24. "Antitrust Shield Mulled for IEP Efforts," *Oil and Gas Journal*, vol. 72, no. 52 (1974), p. 98.
25. Ibid.
26. O. Noreng, "Friends or Fellow Travellers? The relationship of Non-OPEC Exporters with OPEC," *The Journal of Energy and Development*, vol. 4, no. 20 (1979), p. 327.
27. *Agreement on International Energy Program*, p. 11.
28. Ibid., p. 23.
29. Ibid., p. 20.
30. Ibid., pp. 21, 23.
31. "Antitrust Shield Mulled for IEP Efforts," p. 97.
32. "The International Energy Agency, 18 November 1974–18 November 1975," OECD/IEA Press Release, Paris, no date, p. 3. In subsequent years the Board's membership was expanded to approximately thirty companies.
33. "Antitrust Shield Mulled for IEP Efforts," p. 97.
34. Standard Oil (New Jersey) changed its name to Exxon in November 1972. See *Wall Street Journal*, October 25, 1972.
35. "Antitrust Shield Mulled for IEP Efforts," p. 97.
36. *Agreement on International Energy Program,* p. 13.
37. Ibid.
38. Ibid.
39. "Kissinger on Oil, Food, and Trade," p. 67.
40. "The International Energy Agency, 18 November 1974–18 November 1975," p. 1.
41. *Economic Report of the President Transmitted to the Congress January 1978* (U.S. Government Printing Office, Washington, 1978), p. 380.
42. *Energy Balances of* OECD *Countries, 1974/1978* (International Energy Agency and Organisation for Economic Cooperation and Development, Paris, 1980), p. 149.
43. Ibid., p. 150.
44. "OPEC Price-Cutting Protection Sought," *Oil and Gas Journal*, vol. 73, no. 6 (1975), p. 40.
45. Levy, "An Atlantic-Japanese Energy Policy," p. 188.
46. J. M. Blair, *The Control of Oil* (Pantheon Books, New York, 1976), pp. 98–120.
47. *National Institute Economic Review*, no. 68 (1974), p. 87.
48. Ibid., p. 88.
49. Ibid., no. 70 (1974), p. 80.
50. Ibid., p. 82.
51. F. J. Gardner, "Watching the World," *Oil and Gas Journal*, vol. 73, no. 2 (1975), p. 25.
52. *Agreement on International Energy Program*, p. 25.
53. "OPEC Price-Cutting Protection Sought," p. 40.
54. Ibid.
55. "IEA Near Choice of $6–8 as Floor Price," *Oil and Gas Journal*, vol. 73, no. 48 (1975), p. 79.

56. "IEA Rejects Specific Oil Price Floor," *Oil and Gas Journal*, vol. 73, no. 11 (1975), p. 82.
57. "IEA Adopts Safeguard Oil Price of $7/bbl for Oil Imports," *Oil and Gas Journal*, vol. 73, no. 52 (1975), p. 82.
58. *Long-Term Co-operation Programme* (International Energy Agency, Paris, 1976), p. 10.
59. Ibid., p. 11.
60. Ibid., p. 19.
61. Ibid., p. 7.
62. "IEA Near Choice of $6–8 as Floor Price," p. 78.
63. *Long-Term Energy Co-operation Programme*, p. 9.
64. Ibid.
65. Ibid., p. 10.
66. Statistics Canada, *Canada Year Book 1978–79* (Minister of Supply and Services Canada, Ottawa, 1978), p. 898.
67. *Long-Term Energy Co-operation Programme*, p. 14.
68. Ibid., p. 10.
69. Ibid.

CHAPTER TEN

# The Canadian Response

The determination of the United States to regain dominance over oil has had and will continue to have important implications for Canada. It means, among other things, an intensification of the historical attempt to place Canada's energy resources under U.S. control. With unlimited access to Canadian energy, the U.S. can significantly strengthen its position *vis-à-vis* the OPEC countries and the major industrialized powers.

The implications for Canada have been recognized by a number of people. In 1974, for example, James Laxer warned:

> Today, the United States has embarked on the quest for energy self-sufficiency, a new energy strategy that has fundamental implications for the economic and political relations of the western powers. In particular, the new American strategy has vital implications for Canada.... Today, American policy makers intend to achieve self-sufficiency for the United States in part through the inclusion of Canadian resources in the domestic supplies of the United States.... For Canada, the energy crisis is a crisis of industrial policy. The launching of a new American energy strategy confronts Canada with a crucial choice, one that will point this nation's economy either in the direction of further continental integration or toward independence.[1]

Laxer's words were indeed prophetic. The 1973/74 escalation of oil prices and the American reaction to that escalation affected Canada both externally and internally. They forced Canadians to raise basic questions about the use of their energy resources. Among the questions were those relating to the domestic price of energy, the division of the economic rents flowing from higher oil prices among the various sectors of the economy, and the roles of the public and private sectors in the development of Canada's resources.

## The Domestic Price

Immediately following the jump in oil prices, the Canadian government had to decide what its domestic price policy should be. In particular it had to see whether it could fit its two-tier price system, developed under the NOP, into the new, world pricing pattern. As noted in Chapter 7, the NOP prevented Canada from having one integrated national oil market. The market instead was divided in a cartel-like fashion between foreign-supplied and domestic-supplied segments.

One result of this division was that the area east of the Ottawa Valley, which was reserved for foreign oil, received cheaper oil than Ontario and the West, which was the preserve of domestic oil. The price of oil landed in Montreal was based on the posted price of Venezuelan oil, while the price of oil in Toronto was based on the posted price of Alberta oil. Alberta's price was based on the domestic U.S. price, which was higher than world price because of import controls. By 1970, when oil could be purchased in world markets at considerable discounts from the posted price, the discrepancy between the foreign and domestic prices became so large that independent marketers began to ship products from Quebec across the NOP line for sale in Ontario.[2] Imperial Oil estimated that domestic oil in Toronto was 90 cents per barrel higher than imported oil.[3]

At that time compliance with the NOP was voluntary. The movement of foreign oil into Ontario led to a demand by the international oil companies that all imports into Canada be licensed. The National Energy Board (NEB) in 1970 acceded to this demand by licensing all imports and prohibiting the movement of petroleum products across the NOP line.[4] These moves preserved the two-tier price system, much to the delight of the internationals.

### *The End of the* NOP

The NEB's rules became redundant following the escalation of oil prices in 1973, when imported oil became more expensive than domestic oil. Pricewise, there was no longer any danger of foreign oil penetrating home markets. If sufficient transport facilities had existed, like the Edmonton to Montreal pipeline, oil from Alberta and the other western provinces would have captured the entire Canadian market. Since insufficient transportation facilities precluded that from happening, the federal government could have adapted to the new circumstances by continuing the two-price system. However, the continuance of that system would have engendered serious economic and political consequences.

It would have meant that consumers in Montreal would have to pay the high international price for oil while those in Toronto would pay a much lower domestic price. The Toronto price would have been lower as long as it was tied to the frozen U.S. domestic price. Aside from the question of economic justice (which had existed to a far lesser degree when Toronto consumers were forced to pay a higher price than their counterparts in Montreal), there was the question of regional disparities. The continuance of the two-tier price system would have imposed much higher fuel costs on the country's most depressed regions—Quebec and the Maritimes—and would have given the country's most developed province, Ontario, much lower fuel costs. The continuance of the two-tier system would have completely destroyed the policy of reducing regional disparities; it would have made the poor poorer and the rich richer.

At the political level, such a system would have fanned the flames of separatism in Quebec. Only three years prior to the escalation in oil prices, the War Measures Act had been imposed in Quebec in response to the Front de Liberation du Quebec crisis. One can only speculate what might have happened if, in 1974, the Quebecois had been forced to pay much higher oil prices than *les Anglais* in Ontario.

### UNIFIED PRICE

Faced with these economic and political considerations, the federal government had little choice but to abandon the two-tier system in favour of a single Canadian price. This meant that all Canadians would pay the same basic price for oil, with the differences in prices between different locations stemming solely from variations in transportation charges. This establishment of a unified price put an end to the market division imposed by the NOP. It also gave birth to the rudiments of a truly national oil policy. The NOP was, in fact, a misnomer. It should have been more appropriately labelled either IOP (International Oil Policy) or COP (Continental Oil Policy).

The necessity to have a unified price and, as a corollary, an integrated national market raised the important question of the level of that unified price. Was it to be equal to the price of foreign oil landed in Montreal or to the Edmonton price, which was then still linked to the lower U.S. price? The controversy over this question was to dominate the Canadian energy scene for most of the 1970s and during the early part of the 1980s. However, before the price issue is discussed, one other event that helped end the NOP deserves comment.

The Arab oil embargo of 1973/74 threatened most of eastern Canada with an oil shortage. It must be remembered that most of the oil flowing into Montreal is unloaded at Portland, Maine, in the United States and traverses Maine into Quebec. Since the embargo was directed against the United States, it had the potential of affecting unloadings at Portland, placing the Montreal oil supply in jeopardy. This threat, which had been dismissed in hearings before the Borden Commission, was transformed into a reality following the Arab-Israeli war of 1973.

#### EDMONTON-MONTREAL PIPELINE

The turn of events led both the United States and Canadian governments to re-evaluate their attitudes towards a Montreal-Edmonton pipeline. The U.S. government turned from an opponent into an advocate of such a line.[5] The changed international situation led the U.S. to re-emphasize Collado's old policy of having the Western Hemisphere served by Western Hemisphere oil. The U.S. government therefore raised no objections when Trudeau announced to Parliament in December, 1973 that he intended to build a pipeline to Montreal. The construction of this line, he said, was "the single most urgent step toward the attainment of our national goals."[6] The decision to build this line spelled the death knell of the NOP. Canada, at long last, was able to have an integrated national oil market.

### *The Domestic Price Level*

The pipeline, however, did not eliminate the need for imports. Initial pipeline capacity of 250,000 barrels per day[7] accounted for less than one-third of the oil demanded in eastern Canada. Despite the line, most of the East remains heavily dependent on imported oil. This heavy dependence had important implications for domestic oil pricing, and made it very difficult for Canada to have an oil price based on Canadian costs.

That the federal government wanted such a price was indicated in a 1974 speech by Prime Minister Trudeau to the Canadian National Committee of the World Energy Conference. In that speech, Trudeau rejected the world price by asking:

> Is that reasonable price for oil necessarily the highest price charged for oil anywhere in the world? If we argued today, for example, that the world price is reasonable at $10 per barrel of oil, would we argue tomorrow that $20 is reasonable? Or do we see some obligation to consider the cost of producing that energy here in Canada?[8]

The establishment of a price based on domestic production costs would have implied a reduction in price. As mentioned before, the price of Alberta oil was based on the U.S. price, which, in turn, was based on Texas production costs, the highest in the world and considerably above

**Table 10.1: Cost of Finding, Developing, and Producing a Barrel of Conventional Crude Oil, United States and Canada, 1966–70**

| *Costs* | *U.S.* ($U.S. per Barrel) | *Canada* ($U.S. per Barrel) | *% Difference* (U.S./Canada) |
|---|---|---|---|
| Exploration | 0.27 | 0.20 | 35.0 |
| Development Drilling | 0.30 | 0.08 | 275.0 |
| Land | 0.20 | 0.16 | 25.0 |
| Producing Facilities | 0.14 | 0.09 | 55.6 |
| Lifting | 0.53 | 0.27 | 96.3 |
| Total Real Costs | 1.44 | 0.80 | 80.0 |
| Taxes & Royalties | | | |
| Royalties | 0.44 | 0.30 | 46.7 |
| Taxes | 0.16 | 0.14 | 14.3 |
| Total Taxes & Royalties | 0.60 | 0.44 | 36.4 |
| Total Costs | 2.04 | 1.24 | 64.5 |

Source: "Canada has 80 cents advantage," *Oilweek*, vol. 23, no. 1, February 21, 1972, p. 79.

Alberta's costs. A survey by *Oilweek*, for example, revealed that in the period between 1966 and 1970 it cost "80 cents less to find, develop and produce a barrel of oil in Western Canada than the United States, even taking into account the huge new reserves on Alaska's North Slope."[9] This discrepancy can be seen in Table 10.1.

It would seem that, under the NOP, purchasers of western Canadian crude were paying a much higher price in relation to costs than were purchasers of U.S. crude. This was the logical result of linking the Alberta price to the U.S. domestic price.

### THE PROFITABILITY OF OIL

What is even more revealing is that Canadian costs were apparently declining during this period. In their study of costs, Roseman and Wilkinson found that, on the basis of data reported by the Canadian Petroleum Association, "the costs of operating wells (including flow lines and related facilities) decreased 6 cents per barrel over the years 1960–62 to 1969–70 in spite of increasing wage rates and other price increases."[10] They attributed this decline to the change Alberta made in its prorationing system in 1965. This change, which was discussed in Chapter 7, enabled the large producers to utilize their capacity more fully, thereby lowering production costs. It also discouraged over-drilling.

This decline apparently made Canadian oil very profitable. Standard

Oil (New Jersey) recognized this profitability when, in the late 1960s, it recommended that its U.S. affiliate, Humble, import Canadian oil rather than oil from other sources. A 1968 Jersey report, for instance, estimated the average cash flow per barrel of crude, above a specified "cut-off return," at $1.00 for Canadian imports into the U.S. as opposed to 87 cents for imports from its Venezuelan subsidiary, Creole.[11] (According to Jersey, it received a higher cash flow from Creole than from its other overseas sources.)

Jersey's interest in Canadian oil was heightened by the forecasts that U.S. production would decline in the 1970s. It therefore began to explore alternative sources, of which Canadian oil, at that time, was the most profitable. Jersey, in the words of the Bertrand Report, "favoured Canadian imports over alternate crude sources because this source of imports offered [it] the greatest profit—provided that the United States import quota system was not radically changed."[12]

The Bertrand Report noted that Jersey accordingly "formulated a strategy to promote Canadian exports to the United States."[13] It did this by "making certain that new pipelines were constructed from Canada to the United States."[14] It was for this reason that the so-called "Chicago loop" to the IPL was constructed in 1967. This loop enabled Exxon's refineries in the Chicago area to receive Canadian crude.

### RESERVE ESTIMATES

To implement this strategy, Imperial Oil embarked on a campaign to convince the Canadian public that the exports of oil and gas were in their interests. This campaign was necessary because of the growing concern in Canada over the exports of these non-renewable resources. In its 1972 annual report, Imperial stated:

> In the current debate, the export of Canada's energy resources is being questioned; in effect, we are being urged to "bank" our petroleum resources. Our present energy reserves, using present technology, *are sufficient for our requirements for several hundred years*. Export markets will not wait on our convenience. Hydrocarbons can be secured from other geographic sources, and technology leading to the development of other energy forms is being pursued vigorously. In other words, once lost, export markets cannot easily be regained—assuming that they can be regained at all—and their loss would be a genuine economic setback for Canada.[15] (Italics added)

This and similar statements apparently had an impact on government officials. As the Bertrand Report put it, "Imperial's recommendations were not without force."[16] The report cites a 1961 memo by the president of Imperial stating that Imperial had "been confidentially requested to help the NEB devise a discriminatory licensing system."[17] A 1968 Imperial document noted that "Imperial is in somewhat the same position in its relations with Ottawa"[18] as Jersey was in Washington. This document pointed out that, in the words of the Bertrand Report, Jersey "led the way in coordinating and even in writing government policy."[19]

Canadian government officials echoed Imperial's optimism about the huge extent of Canada's resources. In 1971, J. J. Greene, then minister of energy, mines, and resources, told the Canadian Institute of Mining and Metallurgy that Canada's "reserves represent 923 years supply for oil and 392 years for gas" at the 1970 rates of production.[20] Greene, when asked about the sources of information, admitted that the petroleum industry was one of them.[21]

Three years later, on the eve of the energy crisis in 1973, another minister of energy, Donald MacDonald, issued a report which stated that "present indications of Canada's oil and gas potential suggest that there is probably more than enough energy resources to meet domestic requirements until at least the year 2050 *with the possibility of substantial amounts of oil and gas being available for export*"[22] (italics added). These optimistic forecasts, incidentally, were based on the assumption that in the year 2000 the price of oil would be "about $7 or $8 per barrel" and the price of natural gas "about $2 per thousand cubic feet" in terms of 1972 dollars.[23] Presumably if prices would rise more, more oil and gas would be available for export.

As it turned out, world prices rose much faster than the wildest predictions had anticipated. But, lo and behold, Canada's abundant reserves had suddenly vanished. The predictions of both the oil companies and the government turned from elation to despair. What brought about these dramatic changes?

Certain actions by both the U.S. and Canadian governments upset the old cost and price relationships which had made exports so profitable. In 1974 both governments placed controls on the prices of old (*i.e.*, already discovered) domestic oil and prevented them from rising to world levels. There were many reasons for this imposition of controls.

One was political. An overnight, threefold rise in the price of oil in countries that had grown accustomed to cheap, domestically-produced oil could have caused a major governmental crisis. An example of this occurred in 1980 when Prime Minister Joe Clark attempted to raise the price of gasoline by 18 cents a gallon; he was voted out of office.

A second reason concerned the huge economic rents which would have been engendered by a rise in price to world levels. If it was already profitable to produce domestic oil in 1973 at an average price of $3.89 in the U.S.[24] and of approximately $3.50 in Canada,[25] how much more profitable would it have been to produce that same oil at a price of approximately $11 per barrel? A substantial increase in costs was not likely, for the simple reason that the bulk of the costs were "sunk" costs (*i.e.*, costs already incurred in exploration and drilling). Thus a three-fold increase in the domestic price would have given the oil companies huge windfall profits.

The disposal of these economic rents became an important political question in both countries. In Canada, where the provinces are the owners of most sub-surface mineral rights, it became one of the most important issues dominating federal–provincial relations during the 1970s. Because the politics of the economic rents question could not easily be resolved, the governments of the United States and Canada decided to keep the price of domestic oil well below the world price.

But beyond this decision, the pricing policies of the two countries differed. The United States adopted a blended price, consisting of a relatively low price for old oil, higher prices for newly discovered oil, and the world price for imports; the price that the U.S. consumer paid was a composite of these various prices. Canada, on the other hand, opted for a single price completely controlled by the federal government. Unlike in the United States, the Canadian price did not rise either as a result of an increase in the price of imports or as a result of an increase in imports themselves. To maintain this single price, it was necessary to subsidize imports. This, as we shall see, led to serious financial problems.

The differences in policies led to changes in the price differentials between Canadian and U.S. oil. While the price of Canadian domestic oil was higher than that of U.S. old oil, it was nevertheless below the blended U.S. price and considerably below that of world oil. Thus, American refiners had a strong incentive to import Canadian oil.

However, this incentive was destroyed by two events: (1) the imposition by the federal government in January, 1974 of a tax on the exports of crude, and (2) the increases in the royalties charged by the province of Alberta. The export tax was essentially the difference between the Canadian domestic price and the price of overseas oil in the Chicago market. Since this tax equalized the prices of Canadian and overseas oil in Chicago, it destroyed the incentive for purchasers in that area to substitute Canadian for overseas oil. Since it was also a 100 per cent tax on the incremental income of the Canadian subsidiaries of the internationals, and of other potential Canadian exporters, it destroyed their incentive to push for increased exports. This incentive was further decreased by the higher Alberta royalties, which increased the cost of production in the province.

Probably the aspect of the export tax most disturbing to the Canadian oil companies was its use to subsidize imports. In order to insulate the domestic from the world price, the government paid importers in eastern Canada the theoretical difference between the price of foreign oil landed in Montreal and that of Alberta oil transported to that city. This subsidy was paid for from an Import Compensation Fund, money for which was to come from the export tax. Thus any increase in Canadian exports could encourage an increase in subsidized imports. As long as exports remained high, the government felt no financial pressure to reduce import subsidies and, therefore, no pressure to raise domestic prices. If, on the other hand, imports exceeded exports, the revenues of the Import Compensation Fund would be insufficient to pay for subsidized imports. A deficit in the Fund would force the government either (1) to raise taxes to make up the deficit, a politically risky move, or (2) to raise the domestic oil price to world levels, thereby eliminating the need for subsidies.

The oil companies naturally favoured the latter alternative since the tax on the increase in domestic prices would be considerably less than the 100 per cent export tax. In addition, the oil companies were campaigning for decontrol of prices in the United States, and the reduction or elimination of Canadian supplies would place added pressure on the U.S. government to lift controls. The companies therefore felt that increased exports no longer served their interests.

By a remarkable coincidence, their estimates and Canadian government estimates of oil and gas reserves plummeted at the same time. In October 1974, for instance, only sixteen months after MacDonald's *An Energy Policy for Canada* had given assurances that there would be adequate supplies of oil and gas for both domestic and export markets until 2050, the NEB announced that "Canadian oil supplies would be inadequate to serve traditional Canadian markets as well as part of the Montreal area beyond 1982."[26] It recommended that exports be phased out.

This new, much lower estimate is all the more remarkable in light of actual price behaviour between 1972 and 1974. The estimates in *An*

*Energy Policy for Canada* were based on the assumption that the price of oil in 1972 dollars would rise at a rate of approximately 3.7 per cent per year. Given a 1972 price of $2.92 per barrel,[27] this would have implied a 1974 price of approximately $3.14. In actuality the 1974 Alberta price, expressed in 1972 dollars, was $5.23.[28] By 1974 the price had been rising at an annual rate of almost 34 per cent over the 1972 level. Given such unprecedented price increases, one would have anticipated the reserve estimates would have risen sharply; the higher prices would have encouraged increased exploration, which, in turn, would have discovered new reserves. But all that the higher prices engendered was a dramatic drop in reserves.

The paring down of reserve estimates continued well past 1974, even though domestic oil prices continued to rise at greater pace than the rate of inflation. In September 1976, Jim West, senior editor of the *Oil and Gas Journal*, wrote:

> Canada's outlook for crude-oil supply is bleak. And this poses some tough policy decisions for Ottawa.
>
> The Canadian Government began getting the word late last month from some of the country's biggest producers: Western Canada can't much longer supply sufficient crude to its traditional market of the Ottawa Valley, much less meet the government's long-term goal of moving 250,000 b/d [barrels per day] of Western crude to Eastern Canada.
>
> As a result, the flow through a new pipeline extension from Sarnia to Montreal may have to be reversed as early as 1982 to move imported crude to the province of Ontario.[29]

The "biggest producers" to whom West referred were Imperial, Gulf, and Shell.[30] Imperial was the company that four years earlier had stated quite unequivocally: "Our present energy reserves, using present technology, are sufficient for our requirements for several hundred years." It does seem strange that the leading Canadian oil company, which presumably has at its disposal the most modern techniques of exploration and estimation, could make such an enormous error in its reserve figures. The discrepancy raises questions either about the competence of Imperial or about the validity of reserve figures in general. While no one will dispute that it is difficult, if not impossible, to make precise estimates of reserves, one can raise legitimate questions about the extent of the changes in the estimates. Table 10.2 indicates how the estimates of the Geological Survey of Canada changed between the years 1969 and 1976. These estimates, it must be remembered, were based on data furnished to the Geological Survey by the oil companies.

In 1976 the Geological Survey changed its method of estimating reserves. Instead of pinpointing a specific figure, it assigned a probabil-

**Table 10.2: Estimates of Ultimate Recoverable Reserves of Oil and Natural Gas Made by the Geological Survey of Canada, Selected Years**

| *Year* | *Oil*[a] (Billions of Barrels) | *Gas* (Trillion Cubic Feet) |
|---|---|---|
| 1969[b] | 141 | 725 |
| 1972 | 134 | 907 |
| 1973 | 99 | 784 |
| 1976 | 17 (100%) | 100 (100%) |
| | 38 (50%) | 282 (50%) |
| | 80 (0%) | 681 (0%) |
| Ratios, 1976 Highest Estimate (0%) to Lowest (100%) | 8.3 | 9.1 |
| to Most Likely (50%) | 3.7 | 3.2 |

*Note:* Figures in parentheses indicate probability assigned to each estimate.

a. Includes natural gas liquids.

b. Estimate by Canadian Petroleum Association and accepted by National Energy Board.

Source: National Energy Board, *Energy Supply and Demand in Canada and Export Demand for Canadian Energy, 1969–1990* (Ottawa, 1969), pp. 57, 64.

ity estimate to a range of figures. Thus the Survey estimated in 1976 that there was no possibility that ultimate reserves of oil would be as much as 80 billion barrels or that reserves of natural gas would be as much as 681 trillion cubic feet. These top limits were far below the reserve estimates made only three years earlier. The ratios between the highest and lowest, and the highest and most likely, estimates for oil and gas reserves indicate wide discrepancies and uncertainty, which can hardly inspire confidence in the validity of the data furnished to the government by the oil companies.

Nevertheless, the NEB accepted these lower reserve estimates and recommended that exports eventually be phased out. It invoked the licensing power granted to it under Section 87 of the NEB Act to restrict the exports of both crude and products. The subsequent drop in exports became more pronounced than the drop in imports that was supposed to occur as a result of the construction of the oil pipeline to Montreal, despite the presumed substitution of some of the previously exported oil for imports in the Montreal market.

As a result of the more rapid decline in exports, Canada was transformed from a net exporter to a net importer of oil. This change in position had two effects. (1) It forced the Import Compensation Fund, which was supposed to be financed by the export tax, to incur an

**Table 10.3: Canada's Trade Balance in Crude Oil, 1968–81 (Millions of Barrels)**

| *Year* | *Exports* | *Imports* | *Balance* |
|---|---|---|---|
| 1968 | 169.3 | 177.7 | - 8.4 |
| 1969 | 202.7 | 193.4 | 9.3 |
| 1970 | 244.4 | 207.9 | 36.5 |
| 1971 | 270.5 | 245.1 | 25.4 |
| 1972 | 340.9 | 281.6 | 59.3 |
| 1973 | 420.2 | 328.2 | 92.0 |
| 1974 | 333.4 | 291.7 | 41.7 |
| 1975 | 262.7 | 298.6 | - 35.9 |
| 1976 | 182.7 | 277.2 | - 94.5 |
| 1977 | 121.0 | 239.4 | -118.4 |
| 1978 | 98.2 | 231.8 | -133.6 |
| 1979 | 105.2 | 222.4 | -117.2 |
| 1980 | 78.1 | 204.8 | -126.9 |
| 1981 | 59.2 | 195.9 | -136.9 |

Calculated from: Canadian Petroleum Association, *Statistical Year Book*, various years.

increasingly large deficit. (2) It placed pressure on the federal government to raise the domestic price to the world level.

As long as Canada ran an export surplus, it could insulate the domestic price from the world price, if it so desired. Most OPEC nations pursue such a policy by charging a lower price in the the domestic than in the export market. As Table 10.3 shows, Canada was in a position to do likewise up through 1974. Since that year, however, imports have exceeded exports, and Canada has run a consistent deficit in its crude oil trade. This deficit has exceeded 100 million barrels per year for every year since 1976.

These large deficits threatened the federal government's ability to insulate the domestic from the world price. As the deficit started growing, it imposed an increasingly severe financial strain on the federal treasury. A low domestic price could therefore only be maintained by raising taxes or by borrowing. Neither of these options was politically attractive to the federal government. The pressure on the treasury was an important factor in forcing the federal government to agree to raise the domestic price by steps to the world level. Trudeau was forced to abandon his earlier position that there is "some obligation to consider the cost of producing the energy here in Canada." The world price, rather than domestic production costs, was to determine what Canadians had to pay for their oil.

This dependence on imported oil was the inevitable result of the NOP. By making Canada simultaneously dependent on exports for its markets and on imports for its supplies, it destroyed the possibility of allowing Canada to develop an independent oil policy consistent with its own national interests. If the pipeline to Montreal had been built in the 1950s and if the entire Canadian market had been supplied by domestic oil, Canada would never have faced the dilemma of being forced to go to high, world prices because of the need to import a portion of its oil. Canada is the only major exporter of oil that has been *forced* to raise its domestic price to the world level. To the extent that other oil-exporting countries may have raised their domestic prices to the world price, they have done so as a matter of choice. Even if it is desirable, as some economists argue, that Canadian domestic prices rise, they are not rising for sound internal reasons. They are rising because it has become too costly to subsidize imports—imports which would never have been needed had Canada adopted a truly national oil policy when it had the opportunity to do so. Just as in Europe, this dependence on imported oil forced Canada to adopt policies consistent with the interests of the international oil companies.

Though the federal government agreed in 1974 to raise prices to the world level, it could not do so in one fell swoop. It faced a number of political and economic constraints. Probably the most important political constraint was in its dealings with the oil-producing provinces in general, and with Alberta in particular. Since Alberta and the other oil provinces (Saskatchewan and British Columbia) derived a considerable portion of their revenues from the sale of oil and gas, they had an interest in raising prices to the world levels as quickly as possible. In this respect, their interests coincided with those of the oil companies. By acceding to their demands, the federal government would have conceded to them the right to set oil prices. Ottawa, as the government responsible for the entire nation's economic welfare, could not concede that right to any individual province.

## The Disposition of Economic Rents

A related issue concerned the disposition of the economic rents flowing from the price increases. Given the relatively low cost of producing Alberta's oil, any substantial price increase would generate enormous economic rents. Since there is no economic criterion governing how these rents are to be distributed, their actual disposition depends on the bargaining power of the various parties claiming the rents. Rents, it must be remembered, are not costs of production but are a residual item, left over after all the costs of production, including a normal rate of profit, have been paid. While both economic theory and reality dictate that these production costs must be paid if there is to be any output, there is nothing in theory to dictate who should receive the economic rents. They are up for grabs.

Because of the federal nature of the Canadian state, the main parties vying for the economic rents are the provinces, the federal government, and the oil companies. In contrast, in those oil-producing areas which have unitary governments, the chief contenders are the central government and the oil companies. The entry of the provinces into the arena has greatly complicated the problem of disposing of the rents. The federal government was not about to grant the provinces' demands for price increases without coming to some agreement about the sharing of rents. The provinces, as the owners of the resources, were at the same time not willing to acknowledge that the federal government had any legitimate claim to these rents.

The federal government, for its part, wanted the economic rents not only for its own revenue needs but also to retain its control over the country. If the bulk of the rents flowed into the coffers of the provincial treasuries, the economic power of the provinces would grow relative to that of the federal government. It would then become increasingly difficult, if not impossible, to formulate any effective national economic policy. With their increased economic base, the provinces would be in a position to wield far more political power than in the past. This dispersion of power from the centre to the periphery would pose the danger of the balkanization of Canada. In more specific terms, Alberta, as the chief recipient of economic rents, would become the most powerful province in Confederation and would be in a position to undermine federal jurisdiction over the nation's affairs. The federal government was not inclined to abdicate its role, especially to a province containing only one-tenth of the population. Thus, it was essential for the federal government that it obtain a certain minimum share of the economic rents before it consented to large increases in prices.

Another factor preventing a once-and-for-all rise to the world level was the pricing policy of the U.S. government. As has already been pointed out, the U.S. like Canada, prevented domestic prices from rising as rapidly as world prices. The blended U.S. price, while increasing as imports and newly-discovered oil furnished a greater portion of the U.S. supply, was nevertheless below the world price. Even though the traditional links between U.S. and Canadian prices had been drastically weakened, they were not completely broken. For one thing, the U.S. price provided an effective ceiling to the Canadian one.

The relatively weak Canadian manufacturing sector had always had to face strong U.S. competition in both U.S. and Canadian markets. Because of the branch-plant nature of much of manufacturing, firms in Canada could not compete effectively against those in the United States. One of the few advantages Canadians had was access to lower fuel costs. If Canadian oil prices had risen above those of the U.S., this fuel cost advantage would have been lost. Such a development would have spurred the de-industrialization of Canada, reinforcing its traditional role as a hewer of wood and a drawer of water.

*The National Bourgeoisie*

While undermining the position of the manufacturing sector, a rise in oil prices would have greatly strengthened the position of the oil companies. Chapter 11 will clearly point out the implications of this strengthened position for most of the market economies. For Canada, whose oil industry in the mid-1970s was completely dominated by foreign interests, it posed the problem of increased foreign domination of the domestic economy as the cash-rich oil companies began to expand into other fields. Either the oil companies repatriated their profits, thereby causing severe strains on the balance of payments, or they increased investment in the Canadian economy, thereby subjecting more and more sectors to foreign control. Neither prospect was pleasant to the national bourgeoisie, who wanted to free themselves from foreign control.

The national bourgeoisie, furthermore, looked upon oil as a means of gaining its economic independence. Like the rising bourgeoisie in the OPEC countries, they began to see how they could use oil as a means of changing international power relationships in their favour. But to do so they had to wrest the control of oil away from both the provinces and the international oil companies. That this would be no easy task should have been clear from the beginning. In any event it led to protracted, and still continuing, struggles with the oil-producing provinces, with the oil companies, with the U.S. government, and, to a somewhat lesser extent, with the European countries in the Common Market.

The national bourgeoisie, as represented by the federal government, have instituted a number of measures to achieve its goal. These measures have included: control of foreign investment into Canada, direct entry into the oil industry, control of oil prices, and changes in tax laws.

## FOREIGN INVESTMENT REVIEW AGENCY

Among the first measures was the establishment of the Foreign Investment Review Agency (FIRA) in 1974. While, as already mentioned in Chapter 9, this agency screened all foreign investments, it was of special interest to the oil industry because of the large amount of foreign investment in oil. In 1974, for instance, 74 per cent of the assets in minerals were controlled by foreign-owned corporations. The percentage of foreign control in oil was much higher than that in all Canadian industries (22 per cent) or in all non-financial industries (33 per cent). Table 10.4 gives further comparative data.

Given the large book value of foreign investment in non-renewable resources, approximately $15 billion in 1974,[31] any increase in the profits of the resource companies could be used to expand vastly their investments in both resource and non-resource industries. These companies had the theoretical potential of buying out the entire Canadian economy. FIRA, by screening foreign investments, was supposed to prevent this from happening. Regardless of whether FIRA has, in fact, been effective in its screening function, the oil companies have regarded it as an obstacle to their freedom to invest in any industry without encumbrance. The very existence of FIRA has caused, and will continue to cause, conflict between the federal government, the provinces, the oil companies, and the U.S. government. The latter three would like to see FIRA either considerably weakened or abolished altogether. The federal government nevertheless regards FIRA as an essential instrument to enable it to control future energy developments.

**Table 10.4: Assets of Foreign-Controlled Corporations as a Percentage of Industry Assets, 1974**

| *Industry* | *Percentage* |
|---|---|
| Mineral Fuels | 74.0 |
| Total Mining | 63.0 |
| Manufacturing | 56.6 |
| Non-Financial Industries | 32.8 |
| Services | 23.4 |
| Agriculture | 9.8 |
| All Industries | 22.1 |

Source: *Report of the Royal Commission of Corporate Concentration* (Minister of Supply and Services Canada, Ottawa, 1978), pp. 191–92.

PETRO-CANADA

Another important step taken by the federal government was the creation in 1975 of Petro-Canada, a federal Crown corporation. This corporation posed an even greater threat to the foreign-owned oil companies than did FIRA. While FIRA played a passive role, *i.e.*, it restrained the expansion of foreign-owned companies without taking any initiative on its own, Petro-Canada had, and still has, the potential of playing an active role in the energy economy. Petro-Canada represented direct federal government intervention in oil and gas, and could eventually lead to public ownership of the entire industry. It is therefore not surprising that Petro-Canada has been attacked with much greater vehemence than FIRA.

The federal government first toyed with the idea of creating a Crown corporation in oil during the energy crisis of 1973. This option, while not contrary to Canadian tradition, is contrary to the conventional notion of the role of government in a free-enterprise economy. This conventional role assigns to government the task of providing what Adam Smith called "public works," essentially those products or services which, while serving a useful social purpose, are basically unprofitable. Because they are unprofitable private enterprise would not provide them, and therefore government should. Adam Smith gave as his example a lighthouse. In a modern society this concept has been extended to the "infrastructure," *i.e.*, the roads, communications networks, ports, research and development projects, and other facilities needed to attract private industry. Public ownership in Canada has in the past generally been confined to these "infrastructure" industries. It has, by and large, been excluded from the profitable industries.

The establishment of Petro-Canada marked a break with this tradition. The government deliberately entered an industry which had historically been highly profitable and which had prospects of being even more profitable in the future. There were a number of reasons for this dramatic break with traditional free-enterprise values.

First, in practical political terms, Canada was then run by a minority government. The New Democratic Party (NDP), which held the balance of power, had been demanding that the government establish a publicly-owned oil company. But the contention that it was solely NDP pressure that forced the government to set up the Crown corporation is incorrect. As Pratt comments:

> This is a parental claim with little basis in historical fact. The N.D.P. pressures certainly influenced the timing of the decision, yet it seems probable—given the internal bureaucratic pressures operating on the

> Cabinet—that a majority Liberal government would soon have reached the same conclusion (the Liberals certainly did not hesitate to carry through the decision after obtaining a parliamentary majority in the 1974 election).[32]

What was probably far more important than the NDP pressure was the realization by the national bourgeoisie—whose base was largely in manufacturing, trade, finance, publishing, and services—that they were overly dependent on a few large multinational firms for their oil supplies. It was therefore in their interests to reduce this dependency. In addition, the sharp rise in oil prices increased the relative profitability of oil *vis-à-vis* other industries. This increased profitability provided an incentive to enter the oil industry.

But the individual members of the national bourgeoisie were too weak to enter an industry so dominated by the giant firms. Like the bourgeoisie in many other oil-producing and oil-consuming countries, they decided to establish a national company, which would have the financial resources to compete with the giants. This move towards national oil companies is an example of the metamorphosis that capitalism has undergone since it changed from a competitive to a monopolistic environment; there has been an amalgamation of private and public interests. The dominant sectors of the national bourgeoisie have increasingly used the state to further their own specific interests.

Thus the Liberal government, representing the dominant faction of the national bourgeoisie, was able both to appease the NDP and to act in its own interests through the creation of Petro-Canada. In addition, it was able to further both these ends by embarking on a policy of achieving energy "self-sufficiency." This policy involved encouraging the development of indigenous energy resources, like the tar sands in Alberta.

The tar sands, however, posed special problems for the federal government. They were located in an area under provincial jurisdiction; moreover, the province of Alberta had granted leases to private companies to develop the sands in the Syncrude project. There was therefore little scope for direct federal involvement. This situation changed when, in 1974, one of the partners in the Syncrude project pulled out, threatening the viability of the entire project. At this point the federal government entered the scene by offering extremely generous concessions to the remaining private partners—Imperial Oil, Gulf, and Cities Service. In return for its generosity, the federal government received an equity share in the project, a share which was later taken over by Petro-Canada.[33]

The move into the tar sands could have been accomplished differently. The federal government could have prevented the project from failing by taking it over completely, *i.e.*, by nationalizing it. However, such a drastic move would have involved it in a major confrontation with the multinational oil companies, the province of Alberta, and the U.S. government. The federal government had no desire to seek such a confrontation at that time. Once the option of nationalization had been ruled out, it faced the choice of either acceding to the demands of the multinationals or abandoning the project entirely. Given its goal of "self-sufficiency," it could not abandon the project.

Though the particular terms of the Syncrude agreement were attacked by most economic nationalists, including the NDP, they nevertheless gave the government a "foot in the door" in the area of non-conventional oil production. Syncrude illustrates clearly the problems faced by a government which wants to bring energy under its control without nationalizing the entire industry. The government has to act as a partner, and usually a junior partner, with the major oil companies, and this raises the important question of "who controls whom."

In addition to entering the petroleum industry directly through Petro-Canada and the acquisition of shares in Syncrude, the federal government sought to control the economic rents through control of pricing and through limiting the provinces' ability to extract these rents. In 1974, it instituted two measures: the Petroleum Administration Act (PAA), which gave it the power to set prices on oil sold interprovincially; and the disallowment of the federal income tax deductibility previously granted to the oil companies for their royalty payments to the provinces.

### ROYALTY DISALLOWMENT

The royalty disallowment was directed mainly against Saskatchewan, which had imposed a 100 per cent royalty on the oil price increases. The deductibility of this royalty would have significantly reduced the federal income tax liabilities of the oil companies and, therefore, the economic rents flowing to the federal treasury. The disallowment forced both Alberta and Saskatchewan to reduce their royalty rates and make other concessions to the oil companies, since both provinces accepted the oil companies' arguments that the disallowment made their operations unprofitable. Some companies actually ended drilling operations, and an exodus of drilling rigs occurred. Whether this exodus was a bluff is difficult to ascertain. What is important is that, once an area becomes highly dependent on oil, the bargaining power of the oil companies significantly increases and the threat of an exodus can bring to them concessions which they otherwise may not have obtained.

From the point of view of the federal government, the disallowment reduced the ability of the provinces to collect the economic rents created by the price increases. The concessions that the provinces made to the oil companies came out of the provinces' potential economic rents. It was not to Ottawa's interests to have these rents rise too fast.

#### PETROLEUM ADMINISTRATION ACT

The PAA was designed to bolster the federal government's bargaining power in its dealings with Alberta, which had been demanding that oil prices rise to world levels. The PAA gave the national government the power to impose unilaterally any price it desired. By limiting the provinces' ability to raise prices, the PAA also limited their ability to collect the economic rents which would have been created by these higher prices.

After much bickering, Alberta and the federal government came to an agreement, which was accepted by Saskatchewan, regulating oil prices. The price of crude oil was to rise by $1 per barrel on January 1 and July 1 of each year until July 1, 1980, when a new agreement would have to be negotiated, or when it reached 90 per cent of the blended price of oil in the Chicago market. Since the latter price was gradually rising towards the world level, the agreement meant that the Canadian price would eventually be set at 90 per cent of the world price.

### *Conflict among the Bourgeoisie*

This agreement might well have been renewed in 1980 had not an event taken place that disrupted the world oil market—the overthrow of the Shah of Iran and his replacement by Ayatollah Khomeini in 1979. Iranian oil production plummeted and world prices zoomed skyward. The gap between Canadian prices, which had been gradually approaching 90 per cent of the Chicago level, and the world price became wider than ever. The government of Alberta demanded that in any new agreement prices rise at a much faster rate than $1 every six months. At the same time it refused to allow the federal government to claim any increased share of this price increment, arguing that any federal tax on the well-head price of oil represented an infringement on provincial ownership of resources. Alberta suggested that the federal government should obtain increased revenues through adjustments in the corporate income tax. This would have presented the federal government with political problems, since it would have been difficult to raise the income tax on the oil industry alone. For one thing, taxing the oil industry alone may have violated investment tax treaties Canada had signed with the United States and other countries. Ottawa could get at the oil industry only by raising the general level of corporate taxation, thereby further penalizing oil-consuming industries who were already hard hit by the oil price increases.

It is interesting to note that Alberta's conflict with Ottawa was with the Progressive Conservative government of Joe Clark as well as with the Liberal government of Pierre Trudeau. Neither of these governments, representing different factions of the national bourgeoisie, wanted to see too large a portion of the economic rents accruing to Alberta. They both feared that Alberta would then dominate the national economy. Their fears were by no means unjustified. Oil not only transformed Alberta from a rural to an urbanized province, but also created a regional bourgeoisie. Because the province owned most of the sub-surface mineral rights, the economic rents from oil flowed to the provincial treasury in the form of bonuses, lease rentals, and royalties. The newly emerging regional bourgeoisie was in a position to use these "socialized" economic rents for its own interests. The growing oil revenues enhanced the economic and the potential political power of the Alberta bourgeoisie, particularly given the concentration of a significant portion of the economic rents in the Alberta Heritage Savings Trust Fund. Established in 1976, this fund, which was to receive 30 per cent of all the province's resource revenues, had the potential of growing so large as to enable Alberta to buy out most of Canada's key industries. This threat was of serious concern to the established national bourgeoisie which, like any vested interest, did not welcome a newcomer.

It was therefore not surprising that neither the PCs under Joe Clark, generally sympathetic to Alberta's desire for a rapid increase in oil prices, nor the Liberals under Pierre Trudeau, less enthusiastic about a rapid price rise, failed to come to any agreement with Alberta. Neither federal party was willing to let Alberta receive the lion's share of economic rents.

Another, though less obvious, participant in this dispute was what might be called the international bourgeoisie, as represented in Canada by the international oil companies. These companies had a common interest with Alberta in that they wanted the oil prices to rise as rapidly as possible to the world level. While allied to Alberta, their interests were not precisely the same as Alberta's. They wanted the lion's share of the economic rents for themselves. This desire placed them in a potential conflict with both the federal government and the Alberta bourgeoisie. Though ostensibly not taking sides in the Alberta-Ottawa conflict, they were not disinterested observers of it.

The downfall of the Joe Clark government in 1980 brought this three-way conflict to a head. The newly elected Liberal government, which had promised to keep prices low, could not raise oil prices as quickly as Premier Lougheed of Alberta desired. Given federal intransigence on prices, Alberta was unwilling to make concessions on economic rents. Despite vigorous negotiations, the July 1980 deadline came and passed without any price agreement. Alberta nevertheless did not raise prices to the world level. It merely increased them by $2, to $16.75 per barrel in August.[34] The landed price of foreign oil in Montreal at that time was approximately $38 per barrel.[35]

### THE NATIONAL ENERGY PROGRAM

Ottawa did not immediately react to this move. Its turn came several months later, in October, when it brought in the National Energy Program (NEP). The NEP posed a direct challenge to Alberta and the other oil-producing provinces, the foreign-owned oil companies, and the U.S. government. It was the most decisive step the national bourgeoisie had taken to assert its dominance in the field of energy. The objectives of the NEP were: (1) to assert federal control over oil prices; (2) to gain for the federal government a greater share of the economic rents produced by the oil and gas industry; and (3) to "Canadianize" the domestic petroleum industry.

In its price policy the federal government categorically rejected using the world price as a guide. It stated:

> World prices are arbitrary and artificial. They do not reflect conditions of competitive supply and demand, nor the costs of production in Canada or other countries. The Government is determined that the price of Canadian oil will not be linked to world prices, but rather will be "made-in-Canada"—determined on the basis of Canadian circumstances, and the needs of the Canadian economy.[36]

It proposed that the price of conventional crude rise $1 every six months from January 1, 1981 until the end of 1983. Between the beginning of 1984 and the end of 1985, it should rise by $2.25 every six months. Beginning in 1986, it would rise $3.50 every six months "until it reaches its appropriate quality-determined level relative to the oil sands 'reference price'."[37] This "reference price" was set at $38 per barrel, the world price on January 1, 1981, and was to rise annually with either the world price or the Consumer Price Index, whichever was lower.[38] The NEP also provided for higher prices for tertiary enhanced recovery techniques, but these prices were to remain below world levels.

Alberta and the oil industry reacted strongly to this proposed price schedule. It meant, for instance, that Alberta's conventional oil would not reach the 1981 world price level until 1986, some five years later. This implied a loss of billions of dollars of revenues to the province and the industry. The losses would be multiplied to the extent that world prices rose above the 1981 level, a rise which nearly everyone in 1980 believed was certain.

To add insult to injury, the federal government imposed new taxes on the oil and gas industry. Two of these taxes, the Petroleum Compensation Charge and the Natural Gas Tax, were designed to raise the price paid by the consumer at a faster rate than the increases in the well-head price. If this incremental price increase had come about as a result of market forces, the revenues to the oil companies and the royalties to Alberta would both have increased. The federal government instead collected the entire revenue from the price increases engendered by these two taxes.

The Petroleum Compensation Charge, which was imposed at the refinery, was levied to pay refiners a subsidy for imports and for the purchase of oil from Syncrude. That latter oil, as we have already mentioned, was to receive the world price in January 1981. Since the Petroleum Compensation Charge was presumably to be shifted from the refiner to the consumer, the consumer was to pay a blended price, which would rise as long as the world price increased at a faster rate than the domestic price. Thus the NEP had instituted a blended price system at consumer level even though it continued a one-price system at the crude level. This duality meant that neither the oil companies nor the oil-producing provinces were to receive any increased income from these increases in the blended price.

The consequences of the Natural Gas Tax were similar. The rise in the domestic price was to go entirely to Ottawa. From the point of view of the oil companies and the oil-producing provinces, the tax on gas exports would bring even worse results. Because the price of natural gas exported to the U.S. was considerably higher than the domestic prices in both Canada and in the U.S., the companies were running into the problem of finding markets for their exports. They were, in fact, unable to export all the gas for which they had licences. If they tried to shift the tax onto U.S. customers, they would face a further contraction of markets and a consequent loss of revenue. If they did not shift the tax, they also would face a reduction in revenue. Therefore, the oil companies and the oil-producing provinces bitterly opposed the export portion of the tax.

Another measure which promised to cut into the actual earnings of the oil and gas industry was the Petroleum and Gas Revenue Tax (PGRT). Levied at 8 per cent of "net operating revenues related to the production of oil and gas, including income from oil and gas royalty interests,"[39] it was technically neither a well-head tax, to which the province of Alberta so vehemently objected, nor an income tax . A well-head tax would have been based on the gross operating revenues or price; this tax, however, was based on net operating revenues, which were equal to price less operating costs. An income tax would have been based on net income (price minus operating and all other costs); these other costs—such as exploration and development expenditures, capital cost allowances, and interest—could not be deducted from price in calculating the PGRT.

The federal government chose the PGRT instead of an income tax for several reasons. Some of them have already been discussed, like discrimination against the oil industry and possible problems arising from international treaty obligations. Another reason was to be able to tax public sector enterprises and charitable organizations which had invested in the petroleum industry. These enterprises and organizations were exempt from paying federal income taxes but not exempt from the PGRT.

There is little doubt that the PGRT was designed to squeeze both the oil companies and the provinces. The PGRT was for all practical purposes a well-head tax, that is, one based on the price of oil and natural gas. This is because the operating costs, the only costs which are deductible, constitute only a small portion of total costs and are very low in relation to price. Furthermore, the PGRT could not be deducted from the income tax. Finally, it could not be shifted to the consumer because prices were fixed by the federal government.

The federal government was probably correct in its judgement that the profit-swollen oil industry could afford to pay that tax. This was true as long as the tax was on the economic rent. However, the government admitted that there may be instances in which "firms are exposed to hardship due to provincial royalty rates in excess of 50 per cent. In such cases, it would seem reasonable to expect the province in question to adjust its royalties."[40] The federal government thus told the provinces that if the PGRT actually threatened the viability of any oil company, they, rather than the federal government, should redress the situation. This move by the federal government to increase its share of rents at the expense of the provinces was the major factor in the confrontation with Alberta.

While this move had some international implications in that it affected the profitability of the multinational oil companies, it did not nearly have the same international repercussions as the "Canadianization" provisions of the NEP. These provisions posed a direct threat to U.S. control of Canadian oil. The government gave a number of reasons for its insistence on "Canadianization." For instance, it denied that there was any longer a need for importing foreign capital for energy development. The government stated:

> The oil and gas industry, far from drawing in foreign capital, has—since the 1974 oil crisis—been a capital exporter. The industry, in addition to maintaining its normal dividend and interest payments, supported net capital outflows abroad of $2.1 billion in 1975–79. Some of these funds represented a return of capital to foreign owners; other represented new foreign investments by Canadian companies. If dividends and interest payments are added to this total, the total outflow over the period 1975–79 becomes $3.7 billion. . . . In addition, the foreign parents have received fees for technological, operating and managerial services.
>
> Moreover, the prospect is for these capital exports to grow. The continued increase in oil and gas prices that will occur means a further large foreign wealth transfer from Canadians to foreign shareholders. *By ignoring the problem of foreign ownership in the past, Canadians have lost a significant share of the benefits of having a strong resource base. If we fail to act now, Canadians will lose once again.*[41] (Italics added)

The problem was actually much deeper than one of reducing the transfer of wealth abroad. It raised the basic question of whether, under foreign ownership, Canadians could receive the maximum benefits from their rich resource base. While one may question whether the government was more concerned with ensuring that Canadians as whole receive the maximum benefits from resource development or with opening up the oil industry to the national bourgeoisie, the fact remains that large-scale foreign ownership acted as a barrier to the use of Canadian resources for Canadian purposes. That this foreign ownership was—and still is—quite large, there can be little doubt. The government noted:

> Of the top 25 petroleum companies in Canada, 17 are more than 50 per cent foreign owned and foreign controlled, and these 17 account for 72 per cent of Canadian oil and gas sales. *This is a degree of foreign participation that would not be accepted—indeed, simply is not tolerated—by most other oil-producing nations.*

From a public policy perspective, the issue of *control* is equally relevant. Clearly, firms can be effectively controlled by shareholders with far less than 50 per cent of the shares. Thus, even if the oil and gas industry were more than 50 per cent Canadian owned, the industry could still be controlled, to a large degree, by foreign interests.[42] (Italics added)

The reduction in both foreign ownership and control became part and parcel of government policy under the NEP. The stated objectives of the NEP were:

At least 50 per cent Canadian ownership of oil and gas production by 1990;
Canadian control of a significant number of the larger oil and gas firms;
An early increase in the share of the oil and gas sector owned by the government of Canada.[43]

These objectives were to be achieved through changes in the tax laws and through increasing the role of Petro-Canada.

Probably of most immediate importance was the revision in the tax laws. This revision abolished the "earned" depletion allowance on all development expenditures and on exploration expenditures in the provinces. The "earned" depletion provision allowed oil and gas firms to reduce their taxable incomes by an amount equal to one-third of their exploration and development expenditures. However, this allowance could be claimed only by firms and individuals with "resource income" so that, as the government has correctly pointed out, the allowance favoured "those who were already in the industry."[44] Since those in the industry were mainly the foreign-owned firms, the depletion allowance "worked against Canadian ownership objectives."[45] In effect, it discrim-

**Table 10.5: Canadian Ownership Rates and Petroleum Incentive Program Grants Scheduled for 1983 (Per Cent)**

| *Ownership Rates* | *Payments for Exploration* | |
|---|---|---|
| | *Provinces* | *Canada Lands* |
| 0–50 | 0 | 25 |
| 50–75 | 10 | 45 |
| 75+ | 35 | 80 |
| | *Payments for Development* | |
| 0–50 | 0 | 0 |
| 50–75 | 10 | 10 |
| 75+ | 20 | 20 |

Source: Energy Mines and Resources Canada, *The National Energy Program, 1980* (Minister of Supply and Services Canada, Ottawa, 1980), p. 40.

inated against those sections of the Canadian bourgeoisie who wanted to enter the exploration and development phases of the oil industry but who had no resource income. Thus, the tax law provided a disincentive to their entry.

Under the NEP the government has abolished the "earned" depletion allowance, with one exception: it remains in effect for exploration expenditures in the Canada Lands, *i.e.*, lands under federal jurisdiction, like the Northwest Territories. The depletion allowance has been replaced by the Petroleum Incentives Program (PIP), which provides grants to companies on the basis of their degree of Canadian ownership. The proposed relationship between degree of ownership and size of grants is shown in Table 10.5.

The table shows that, for example, a Canadian-controlled firm with over 75 per cent Canadian ownership will receive a PIP grant of 80 per cent of its exploration expenditures in the Canada Lands. A foreign-controlled firm, *i.e.*, one with less than 50 per cent Canadian ownership, will receive a PIP grant for only 25 per cent of its exploration expenditures in the Canada Lands, and nothing for its efforts in the provinces. Similar relationships prevail for development expenditures.

The substitution of the PIP grants for the depletion allowance[46] had a number of important implications. First, these grants provide a clear and unambiguous record of payments by the Canadian government to the oil companies. Unlike the "tax expenditures" of the depletion allowance, these explicit payments are subject to public scrutiny. Because of the confidentiality provision in the income tax law, claims for depletion allowances are not subject to such scrutiny.

Second, the system was designed to give a greater incentive for exploration in areas under federal rather than provincial control. As can be seen in Table 10.5, the grants are much greater for exploration efforts in the Canada Lands. To the extent that these grants result in successful exploratory activity in the Canada Lands, they will strengthen the bargaining power of the federal government *vis-à-vis* the provinces.

Third, the new tax regime was designed to open up the oil industry to Canadian investors. The government explicitly stated that the PIP grants were "structured to encourage investment by Canadian companies and individuals."[47] No one had to have resource income in order to benefit from the new tax laws. The petroleum industry was thrown wide open to newcomers. This particular provision was of considerable benefit to those sections of the national bourgeoisie who had previously not invested in the oil industry and now wanted to take advantage of its profit potential.

Fourth, the new tax provisions raised the after-tax earnings of Canadian-owned companies relative to foreign-owned ones. A Canadian-owned firm with a taxable income would receive 93 cents, in PIP grants and tax write-offs, for every dollar it spent on exploration in the Canada Lands. A foreign-owned firm, on the other hand, would receive only 78 cents. The resulting differential in after-tax earnings provided an incentive for foreign-owned firms to "Canadianize" the exploration divisions of their companies by opening them up to Canadian investors. It also provided an incentive for Canadian firms to buy out foreign-owned ones. The mergers induced by this differential in after-tax earnings seriously strained relations between the United States and Canada. The U.S. viewed the mergers as a direct threat to American ownership and control of Canada's oil resources.

Another aspect of the "Canadianization" policy which enraged the U.S. was the increased role assigned to Petro-Canada. This corporation was to receive "a 25 per cent interest in every right in Canada Lands."[48] Together with private Canadian firms, it was to receive at least a 50 per cent interest in " any production from Canada Lands."[49] In addition to this "back-in" right on Canada Lands, Petro-Canada, or one of its agents, was instructed to acquire "several of the large oil and gas firms."[50] These acquisitions were to be financed through a special Canadian Ownership Account, whose revenue would be derived from "special charges on all oil and gas consumption in Canada."[51]

The impact of this and the other provisions of the NEP was enormous. Howls of protest went up from the international oil companies, the United States and other foreign governments, and the oil-producing provinces.

### THE OIL WAR

Following the announcement of the NEP both the Alberta government and the oil industry declared war on the federal government. For its part the industry moved rigs out of Canada into the United States, presenting this move as solely a response to the NEP, when in fact it would have occurred anyway (at least partially), because there was a surplus of rigs in Alberta; in fact, rigs were leaving even before the NEP was announced. The Alberta government's response was to institute a series of production cutbacks. In addition, Premier Lougheed put a hold on two scheduled mega-projects, the Alsands tar sands plant near Fort McMurray and the Esso Resources heavy oil unit at Cold Lake. Both these projects were viewed as key ingredients in the federal government's drive for energy self-sufficiency by 1990.

These measures were not intended to force eastern Canadians to "freeze in the dark." Even if Lougheed had this aim, which is highly doubtful, he could not have achieved it. At the end of 1980, there was plenty of oil available in the world markets. What Premier Lougheed wanted was to force the federal government to increase imports, which would have compelled Ottawa to pay out larger subsidies to eastern refineries, thereby imposing a financial strain which, the government of Alberta believed, would force the federal government to raise prices more rapidly to world levels.

If Lougheed had used this strategy one or two years earlier, he probably would have won a clear victory over the federal government. But when he did institute these cutbacks, the Canadian economy was beginning to falter. The total demand for oil, both domestic and imported, started to fall. As Table 10.3 shows, imports in 1981 continued to fall from previous levels. The hoped-for increase in imports did not materialize.

### THE PEACE AGREEMENT

At the same time the Alberta economy began to suffer from the cutbacks. People servicing the rigs began putting pressure on the premier to end the impasse. Residents of the Fort McMurray and Cold Lake areas were

**Table 10.6: A Comparison of Scheduled Prices for Conventional Old Oil under the NEP and under the Canada-Alberta Energy Agreement, 1982–86**

| *Date* | *NEP* ($ per Barrel) | *Agreement* ($ per Barrel) |
|---|---|---|
| January 1, 1982 | 19.75 | 23.50 |
| July 1, 1982 | 20.75 | 25.75 |
| January 1, 1983 | 21.75 | 29.75 |
| July 1, 1983 | 22.75 | 33.75 |
| January 1, 1984 | 25.00 | 37.75 |
| July 1, 1984 | 27.25 | 41.75 |
| January 1, 1985 | 29.50 | 45.75 |
| July 1, 1985 | 31.75 | 49.75 |
| January 1, 1986 | 35.25 | 53.75 |
| July 1, 1986 | 38.75 | 57.75 |

Sources: Energy, Mines and Resources Canada, *The National Energy Program, 1980* (Minister of Supply and Services Canada, Ottawa, 1980), p. 26, and "Memorandum of Agreement between the Government of Canada and the Government of Alberta relating to Energy Pricing and Taxation," September 1, 1981, p. 2.

eager to have the mega-projects resume. Given the downturn in the Canadian economy and internal political pressures in the province, Alberta had no choice but to accept the main provisions of the NEP. There is little doubt that Alberta was the loser in the agreement it came to with the federal government in September 1981.

What Alberta won were concessions on prices and on the export portion of the excise tax on natural gas. As can be seen in Table 10.6, price increases for conventional old oil were rescheduled to rise at a much faster rate than under the NEP. However, these scheduled increases were based on the assumption that world oil prices would rise at a constant rate of 13 per cent per year. How this 13 per cent figure was arrived at was never fully explained; presumably it was an extrapolation of past trends, and it generated an unrealistic expectation of future price increases in Canada. All prices under the Canada-Alberta agreement were "subject to the condition that the overall average field price of conventional old oil plus transportation costs to Montreal, adjusted for quality, will not exceed 75 per cent of the actual international price of oil."[52]

This 75 per cent limitation turned the Alberta "victory" on pricing into a paper one. Since the agreement was signed in September 1981, world prices, instead of rising at an annual rate of 13 per cent, have actually declined. Barring a major upheaval in the Middle East, like the overthrow of the Saudi Arabian government, it is highly unlikely that the price for Alberta's conventional old oil will rise much above the levels of late 1982 or early 1983. A similar pattern can be forecast for other types of oil, like conventional new oil, synthetic oil, and oil produced by enhanced recovery methods. Their price was to rise immediately to the world level and increase by 13 per cent per year according to the agreement's price schedule. Since the initial rise to the world level, they have remained well below the schedule.

The Alberta victory on the taxing of natural gas was also a hollow one. Instead of abolishing the export portion of the natural gas excise tax, the federal government merely agreed to reduce the rate to zero, "without prejudice to the Government of Canada's position that it has the right to levy such a tax"[53] if it so desires. The government has consented to a zero rate only for the five-year term of the agreement. Upon the agreement's expiration, it has the right to raise the levy above zero per cent. In return for this temporary zero rate, the province undertook to pay for the PIP grants within Alberta, to compensate Ottawa for revenue lost from the zero rate on the export tax.

**Table 10.7: Estimated Distribution of Revenues from Oil and Gas Production (Per Cent)**

| | *1979* | *NEP 1983* | *Agreement 1981–86* |
|---|---|---|---|
| Federal Share | 12.0 | 27.4 | 25.5 |
| Industry Share | 39.5 | 31.6 | 44.3 |
| Alberta's Share | 48.5 | 41.0 | 30.2 |

Calculated from: Energy, Mines and Resources Canada, *The National Energy Program 1980* (Minister of Supply and Services Canada, Ottawa, 1980), p. 108, and "Memorandum of Agreement between the Government of Canada and the Government of Alberta relating to Energy Pricing and Taxation," September 1, 1981, p. 22.

In the key area of rent-sharing the province was the heavy loser. As Table 10.7 shows, the province's share of revenues will decline from almost one-half in 1979 to less than one-third during the years 1981–86, when the NEP had allocated to it a two-fifths share.

Table 10.7 also shows that Ottawa gained considerably over 1979 but lost slightly in comparison to the NEP. The big winner was the oil industry, which gained relative to both 1979 and the NEP. The basis for this oil "victory" was the much more rapid escalation in prices and the reduction of the natural gas export tax. The figures in Table 10.7 must, however, be viewed with caution. They are based on the assumption that oil prices will rise as stipulated under the NEP and in the Canada-Alberta agreement. As mentioned earlier, this seems to be a highly improbable assumption. Also, some academic economists, using a different methodology, would question whether the oil industry's position had improved. But there is general agreement that the position of Alberta had weakened.

This weakening was accentuated by concessions Alberta made to the industry after the agreement was signed. Alberta's concessions were followed by ones from the federal government. These moves have resulted in further increasing the industry's share of revenues.

## The Economic Downturn

The Canada-Alberta energy agreement and the subsequent concessions to the oil industry, while ending the polemics between the province and the federal government, did not revive the depressed oil industry.

Contrary to expectations, conventional oil output did not zoom upward once Alberta removed its production controls. What is even more important, the two large mega-projects, which Lougheed tried to use as an ace in his bargaining with Ottawa, did not come into being. Though Lougheed gave the companies involved permission to resume construction, they refused to do so. Within months after the signing of the agreement, the projects were abandoned, and this struck a heavy blow to Alberta's economy.

The failure of the conventional oil industry to revive and the abandonment of the mega-projects resulted from the severely deteriorating economic environment which Canada faced in the period following the signing of the agreement. The economic downturn reduced the demand for oil, and also put downward pressure on world oil prices. Accompanying this fall in prices was a sharp rise in interest rates. This combination of falling prices and rising interest rates put a damper on investment in the oil industry. The boom, at least for the time being, was over.

## Canadianization

With the end of the boom, the position of the oil industry has actually strengthened. It has called for the abolition of the most onerous aspects of the NEP and has been receiving a friendly ear from large segments of Canadian society. The provincial premiers, from both oil-producing and oil-consuming provinces, have echoed the companies by demanding either the weakening or the elimination of FIRA. There have also been calls to improve the climate for foreign investment, which implies granting greater concessions to the oil industry. To what extent this counter-offensive will seriously weaken or destroy the NEP remains to be seen.

While the NEP may have brought disappointing financial results to the federal government, it has succeeded in one important aspect: it brought about a significant "Canadianization" of the oil industry. Between October 1980 and May 1982, it increased the share of Canadian ownership by almost seven percentage points and the share of Canadian control by nearly eleven percentage points.[54] This rise has been brought about essentially by the activities of two companies, Petro-Canada and Dome Petroleum.

### *Petro-Canada*

Petro-Canada, which prior to the NEP had acquired ARCAN and Pacific Petroleum, purchased Petrofina in 1981[55] and was in the process of buying the marketing and refining facilities of British Petroleum towards the end of 1982.[56] By the end of 1981, Petro-Canada was the fourth largest oil and gas producing firm in the country. In 1979, it had been seventh.[57] The purchase of Petrofina[58] for the seemingly high price of $1.5 billion gave Petro-Canada both producing properties in the West and refining and marketing facilities in the East. The purchase of British Petroleum, which appears to be attractive to the stockholders of that company, will make Petro-Canada the largest marketer in Ontario and Quebec. It will also increase its refining facilities in the East.

These two purchases do raise questions about the cost of acquisitions. Since the NEP does not countenance forced sales in the public interest, the only way Petro-Canada can acquire foreign-owned companies is to offer the potential sellers a highly attractive price. In this classic example of a very willing buyer, with almost unlimited resources, confronting a not-too-eager seller, the seller is bound to get the better of the bargain. The unlimited resources of Petro-Canada stem from its ability to raise funds from the federal government's Canadian Ownership Account. The existence of this account, financed by taxes on oil and gas, also gives it access to other forms of financing; the chartered banks, for instance, are reportedly financing the take-over of BP. The ability of the sellers to extract a high price from Petro-Canada reduces the potential benefits that "Canadianization" can bring to the country.

### *Dome Petroleum*

Dome Petroleum, the other major actor in the "Canadianization" scene, acquired Hudson's Bay Oil and Gas in 1981 for $2 billion. With this acquisition, it became the seventh largest oil and gas producing company in the country, up from twelfth place in 1979.[59] Dome relied on the Canadian chartered banks and banks in the U.S. to finance its purchases. With the economic downturn compounded by high interest rates, Dome ran into serious financial difficulties and, in effect, became bankrupt. It was technically saved by a rescue operation conducted jointly by the chartered banks and the federal government, as a result of which both these groups have obtained control of the company from its shareholders.

The plight of Dome illustrates a potential weakness of the NEP: the reliance on private enterprise to accomplish "Canadianization." The incentive is the differential earnings between Canadian and foreign-owned enterprises. But this type of incentive encourages highly speculative and risky take-overs, of which Dome is a prime example.

*The U.S. Reaction*

Given the experiences of Petro-Canada and Dome, it would seem that "Canadianization" has been successful, but at an extremely high cost. The success of the "Canadianization" program can be gauged by the vehemence of the U.S. reaction towards it. Dome's acquisition of Hudson's Bay Oil and Gas, which had previously been controlled by Continental Oil, aroused concern among U.S. oil companies in Canada. They forced the introduction of legislation into the U.S. Congress demanding protection against what they called "Canadian buccaneering."[60] These companies were represented by George Ball, a former under-secretary of state. The U.S. government sent a sharp and extremely hostile note to Canada, demanding that Canada virtually abandon the NEP.[61] The *Globe and Mail* reported that an unnamed State Department official, who complained that the NEP "is a very large and massive retreat from the principles of national treatment," threatened retaliation by pointing out that there are "many examples of significant Canadian investment in the U.S. which could be taxed, regulated or legally constrained, one way or another."[62]

There is thus an interesting replay of history going on. The dispute with Canada over non-discriminatory treatment of U.S. companies and over "reciprocity" is strikingly similar to the disputes with Great Britain and Holland in the years following World War I. Regardless of how this dispute will be resolved, the United States can no longer be certain that Canada's energy resources are "secure." The NEP is the culmination of a trend that has been growing in Canada for the past two decades, the trend towards economic nationalism. It reflects, in part, the desire of a portion of the Canadian bourgeoisie to use Canada's resources for their own development rather than for the furtherance of U.S. energy policy.

This conflict over the control of resources will continue far into the future. At this writing, it cannot be predicted how this conflict will be resolved. Whatever does happen, the issues have relevance far beyond North America. Basically, they involve the question of whether energy should continue to be used as a weapon in a power struggle, or whether it should be used to foster rational economic development and national independence. This will become the most important issue in a world in which oil will give way to other forms of energy, bringing with it changes in world power relationships and an era of "economic warfare" for control over the new energy sources.

## Notes

1. J. Laxer, *Canada's Energy Crisis* (James Lewis & Samuel, Toronto, 1974), p. 2.
2. Canada, Director of Investigation and Research, Combines Investigation Act, *The State of Competition in the Canadian Petroleum Industry, Findings, Issues and Remedies* (7 vols., Minister of Supply and Services Canada, Ottawa, 1981), vol. 1, p. 16.
3. Ibid., *The Domestic Sector: An Overview of the Environment, Industry Behaviour and Performance*, vol. 2, p. 34.
4. Ibid., *Findings, Issues and Remedies*, vol. 1, p. 16.
5. Laxer, *Canada's Energy Crisis*, p. 85.
6. House of Commons Debate, December 6, 1973.
7. *Canada Yearbook, 1978–79* (Minister of Industry, Trade and Commerce, Ottawa, 1978), p. 567.
8. W. Scotland, "The Federal Government's View of the Canadian Energy Situation" in *Canadian National Energy Forum* (Canadian National Committee—World Energy Conference, 1974), pp. 97–98.
9. "Canada Has 80¢ Advantage," *Oilweek*, vol. 23, no. 1, February 21, 1972, p. 79.
10. F. Roseman & B. Wilkinson, "Who benefits? The Alberta Energy Price Increases," *The Canadian Forum*, vol. 53, no. 629–30 (1973), p. 48.
11. Canada, Director of Investigation and Research, Combines Investigation Act, *The State of Competition, The Domestic Sector*, vol. 2, p. 32.
12. Ibid., pp. 30–31.
13. Ibid., p. 31.
14. Ibid.
15. Imperial Oil, *Annual Report*, 1972.
16. Canada, Director of Investigation and Research, Combines Investigation Act, *The State of Competition, The Domestic Sector*, vol. 2, p. 33.
17. Ibid.
18. Ibid.
19. Ibid.
20. B. Willson, "An Alternative Petroleum Policy for Canada" in J. Laxer & A. Martin (eds.), *The Big Tough Expensive Job, Imperial Oil and the Canadian Economy* (Press Porcepic, Toronto, 1976), p. 198.
21. Ibid.
22. Canada, The Minister of Energy, Mines and Resources, *An Energy Policy for Canada, Phase 1–Analysis, Summary of Analysis* (Information Canada, Ottawa, 1973), p. 14.
23. Ibid.
24. "Analysts See More Gains in U.S. Oil Profits," *Oil and Gas Journal*, November 10, 1980, p. 160.
25. Calculated from: Alberta Energy Resource Conservation Board, file III P-1-1.
26. Canada, The Minister of Energy, Mines and Resources, *An Energy Strategy for Canada: Policies for Self-Reliance* (Minister of Supply and Services Canada, Ottawa, 1976), p. 14.

27. Alberta Energy Resource Conservation Board, file III P-1-1.
28. Ibid. The actual price in 1974 dollars was $6.58. The 1972-dollars figure is arrived at by deflating this by the GNE Implicit Price Index.
29. J. West, "Canada's Domestic Crude-Supply Outlook Seen Grim," *Oil and Gas Journal*, September 27, 1976, p. 19.
30. Ibid., p. 20.
31. *Report of the Royal Commission on Corporate Concentration* (Minister of Supply and Services Canada, Ottawa, 1978), p. 193.
32. Larry Pratt, "Petro-Canada" in A. Tupper & G. Doern (eds.), *Public Corporations and Public Policy in Canada* (The Institute for Research on Public Policy, Montreal, 1981), pp. 95–148.
33. Ibid., p. 20. For an excellent analysis of the tar sands settlement, see L. Pratt, *The Tar Sands: Syncrude and the Politics of Oil* (Hurtig, Edmonton, 1976).
34. Energy, Mines and Resources Canada, *The National Energy Program 1980* (Minister of Supply and Services Canada, Ottawa, 1980), p. 26.
35. Ibid., p. 25.
36. Ibid., p. 27.
37. Ibid.
38. Ibid., p. 28.
39. Ibid., p. 38.
40. Ibid.
41. Ibid., pp. 17–18.
42. Ibid., p. 19.
43. Ibid., p. 49.
44. Ibid., p. 19.
45. Ibid.
46. As mentioned on p. 239, the $33\frac{1}{3}$ per cent earned depletion allowance was not abolished for exploration expenditures in the Canada Lands. It was, however, limited to only those expenditures not reimbursed by PIP grants. If, for instance, a Canadian-owned firm spent $1,000,000 on exploration, it would receive $800,000 in PIP grants. The depletion allowance would then be applied to the remaining $200,000.
47. Energy, Mines and Resources Canada, *The National Energy Program 1980*, p. 39.
48. Ibid., p. 47.
49. Ibid.
50. Ibid., p. 51.
51. Ibid.
52. "Memorandum of Agreement Between the Government of Canada and the Government of Alberta Relating to Energy Pricing and Taxation," September 1, 1981, p. 1.
53. Ibid., p. 9.
54. Minister of Energy, Mines and Resources Canada, *The National Energy Program, Update 1982* (Minister of Supply and Services Canada, Ottawa, 1982), p. 48.
55. Ibid.
56. *Energy Analects*, November 5, 1982.

57. Minister of Energy, Mines and Resources Canada, *The National Energy Program, Update 1982*, p. 49.
58. Ibid., p. 48.
59. Ibid., p. 49.
60. *Globe and Mail*, June 20, 1981.
61. U.S. Dept. of State, "Note on National Energy Program to the Government of Canada," March 5, 1981.
62. *Globe and Mail*, June 22, 1981.

CHAPTER ELEVEN

# The Drive to Regain Dominance

The investment and MSP provisions of the IEA agreements are the vehicles through which the United States will attempt to regain dominance. They provide for the free flow of energy investments among the member states and for a guaranteed rate of return on these investments. Most of these investments will come from American companies, which are controlled by a few large financial groups, the most prominent of which are the Rockefellers. The free flow of investment will enable these financial groups to gain more and more control over the economies of the major consuming countries. This control will be reinforced if these investments turn out to be highly profitable ones, since the profits will be used to buy out companies in a wide variety of industries throughout the world.

## Oil Companies' Profits

Despite OPEC's success in obtaining a much higher share of oil revenues, the profits of the companies have zoomed upward. One reason is that the companies' earnings per barrel of crude increased as the posted price rose. In 1970, when the posted price of Saudi Arabian crude was $1.80, ARAMCO paid 96.3 cents in royalties and income taxes to the Saudi government. The government take was equal to 53.5 per cent of the price. By the end of 1974, when the posted price was $11.25, the Saudi government received from royalties, income taxes, and the sale of its own oil to ARAMCO approximately $10.289, equivalent to 91 per cent of the price. The per barrel earnings of ARAMCO nevertheless increased from 74 cents to 90 cents, or by approximately 22 per cent. These data are shown in Table 11.1.

Since most of ARAMCO's oil went to the affiliates of its owners, the posted price represented merely a bookkeeping transaction, which allocated a company's total profits and costs among its various divisions. The increase in product prices, which represent arm's-length transac-

**Table 11.1: Estimated Distribution of Revenue Received from Sale of One Barrel of Saudi Arabian Oil by ARAMCO, 1970 and 1974**

| | *August 31, 1970* | | *November 1, 1974* | |
|---|---|---|---|---|
| | ($) | (%) | ($) | (%) |
| Posted Price | 1.800 | 100 | 11.250 | 100 |
| Government Revenues | 0.963 | 53.5 | 10.289[a] | 91.5 |
| Operating Costs | 0.100 | 5.6 | 0.065[b] | 0.6 |
| Company Earnings | 0.737 | 40.9 | 0.896 | 8.0 |
| Per Cent Increase, Company's Earnings, 1970–74 | | | | 21.6 |

a. Includes the revenue received by the government from the sale of its own crude to ARAMCO. The company paid $10.665 a barrel for this crude. The figure of $10.289 in the table is a weighted average of the royalty and income tax payments of the company on its own crude plus the company's payments for the government crude.
b. This is the estimated operating costs of producing the company's crude. It does not include the operating costs of the government's crude.

Source: *International Petroleum Encyclopedia 1975* (The Petroleum Publishing Company, Tulsa, Oklahoma, 1975), p. 7.

tions to third parties, was the main reason for the increase in the companies' net earnings. Though most consuming countries imposed controls on product prices, they permitted the companies to pass on their increased costs to the final consumer. Product prices thus rose as the companies increased their payments to the host countries. Moreover, the companies raised their prices much more than their increased payments to the oil governments. Table 11.2 indicates the extent to which the increases in product prices exceeded those of crude. As can be seen, changes in product prices were greater than changes in crude oil prices in every year since 1971, the first year that the price of Saudi Arabian light crude rose from its traditional level of $1.80 per barrel. Over this nine-year period, the estimated increase in the revenues received by the ARAMCO partners came to more than $34 per barrel of refined products sold, while the increases in the payments for the crude oil acquired from Saudi Arabia, the marker crude which roughly reflected the movement in world crude prices, amounted to only $15.50 per barrel. The price of products during this period thus rose by $19 per barrel more than the price of crude. In some years, like 1974,1975, 1978, and 1979, product

**Table 11.2: A Comparison of ARAMCO[a] and Saudi Arabian[b] Price Increases, 1971–79 ($ per Barrel)**

| *Year* | *Product Price Increase* | *Crude Price Increase* | *Difference* |
|---|---|---|---|
| 1971 | 0.864 | 0.395 | 0.469 |
| 1972 | 0.462 | 0.274 | 0.188 |
| 1973 | 1.680 | 0.833 | 0.847 |
| 1974 | 10.319 | 7.538 | 2.781 |
| 1975 | 3.776 | 0.156 | 3.620 |
| 1976 | 1.981 | 0.464 | 1.517 |
| 1977 | 2.662 | 0.930 | 1.732 |
| 1978 | 2.528 | 0.310 | 2.218 |
| 1979 | 10.190 | 4.600 | 5.590 |
| 1971–79 | 34.462 | 15.500 | 18.962 |

a. Estimated increases in price per barrel of refined products received by ARAMCO's private owners—Exxon, Standard Oil of California, Texaco, and Mobil.
b. Increases in the average posted price per barrel of Saudi Arabian light crude.

Sources: Product prices—based on refinery product sales and gross operating data found in *International Petroleum Encyclopedia*, various issues; crude prices—M. Mauksch, P. Carrewyn & C. Redman, *Energy and Europe, EEC Energy Policy and Economy in the Context of the World Energy Crisis* (2 vols., Agence Europeene d'Information, Brussels, 1975), vol. 1, p. 117, and F. Al-Chalabi, *OPEC and the International Oil Industry, A Changing Structure* (Oxford University Press, Oxford, 1980), pp. 85–90.

prices grew much more than crude acquisition costs. The differential in favour of products ranged from $2.22 in 1978 to $5.59 in 1979. While these estimates, based on the relationship between gross operating revenues and refined product sales, are admittedly rough, they are nevertheless indicative of trends. Furthermore, these trends are consistent with changes in the profit position of the oil companies.

A study by Chase Manhattan Bank of the earnings of twenty-six leading oil companies reveals that their net income after taxes rose from $7 billion in 1970 to almost $32 billion in 1979, or by 377 per cent. Furthermore, their after-tax rate of return on invested capital rose from 10.7 per cent in 1970 to 24 per cent in 1979. The average rate of return for the years 1968–72 was 10.7 per cent. In the years 1973–79, the years of significant oil price increases, the average rose to 16 per cent,[1] a level 50 per cent higher than in pre-escalation years. The companies thus increased both the absolute level of their earnings and the level relative to their invested capital. They could not have made these spectacular increases in their rates of return if they had not increased their product prices more than their crude acquisition costs.

In addition, the companies were able to increase their depreciation, depletion, and amortization write-offs considerably. As Table 11.3 shows, these write-offs increased from $7 billion in 1970 to $19 billion in 1979, or by 189 per cent. These write-offs together with their net income constitute the after-tax cash flow received by the companies. This sum is the theoretical, internally generated investible capital available to the firms. As can be seen, the cash flow rose from $13 billion in

**Table 11.3: After-Tax Net Income, Write-Offs, and Cash Flow, 26 Major Oil Companies,[a] 1970–79 ($ Billions)**

| *Year* | *Net Income* (1) | *Write-Offs*[b] (2) | *Cash Flow* (1) + (2) |
|---|---|---|---|
| 1970 | 6.6 | 6.6 | 13.2 |
| 1971 | 7.3 | 7.1 | 14.4 |
| 1972 | 6.9 | 7.5 | 14.4 |
| 1973 | 11.7 | 8.3 | 20.0 |
| 1974 | 16.4 | 10.1 | 26.5 |
| 1975 | 11.5 | 11.2 | 22.7 |
| 1976 | 13.1 | 11.2 | 24.3 |
| 1977 | 14.4 | 13.0 | 27.4 |
| 1978 | 15.0 | 15.6 | 30.6 |
| 1979 | 31.5 | 19.1 | 50.6 |
| Total 1970–79 | 134.4 | 109.7 | 244.1 |
| Per Cent Change 1970–79 | 377 | 189 | 283 |

a. The twenty-six companies are: Amerada Hess, Ashland Oil, Atlantic Richfield, British Petroleum, Champlin Petroleum, Cities Service, Clark Oil and Refining, Compagnie Francaise des Petroles, Conoco, Exxon, Getty Oil, Gulf, Louisiana Land and Exploration, Marathon Oil, Mobil Oil, Murphy Oil, Petrofina, Phillips Petroleum, Royal Dutch Shell Group, Standard Oil of California, Standard Oil (Indiana), Sun, Superior Oil, Texaco, Tosco, and Union Oil. All but four of these companies—British Petroleum, Compagnie Francaise des Petroles, Petrofina, and Royal Dutch Shell—are American.
b. Depreciation, depletion, and amortization.

Source: P. Keenan, R. Dobias, M. O'Neill & N. Anderson, *Financial Analysis of a Group of Petroleum Companies, 1979* (Chase Manhattan Bank, New York, 1980), pp. 2, 22–23.

1970 to $51 billion in 1979. This latter sum, received by only twenty-six companies, was greater than the Gross Domestic Product of most countries in the world.

## Expansion into Other Fields

More important, the investible funds available to these companies were increasing at a time when firms in other industries were encountering financial difficulties as a result of the worldwide recession following the oil price increases. This increase in the relative profitability of the oil sector enabled oil companies to buy out companies in other sectors. The oil companies began to acquire an ever larger share of the ownership and control of the economies of the advanced developed countries. The Rockefellers are a prime example of this process.

### *Rockefeller Group*

This expansion into other fields actually started long before the price rises of the early 1970s. Rockefeller started it on a personal basis in the late nineteenth century. In the 1890s he had sixty-seven major investments in non-oil industries, such as railroads, mining, and banks. Over the years he acquired a controlling interest in Consolidated Coal and in Colorado Fuel and Iron, and invested in such diverse companies as International Harvester and General Motors.[2] At one time he was the largest single shareholder in United States Steel.[3]

#### CHASE MANHATTAN BANK

As Rockefeller grew older, he turned control of his holdings over to his son, John D. Rockefeller, Jr., and to other members of his family. The Rockefeller family, as a group, expanded their interests during the 1930s. In 1930 the Rockefeller group made one of its most important purchases by acquiring between 4 and 5 per cent of the stock of Chase Manhattan Bank, enough to give the family effective control over that bank,[4] which is now headed by David Rockefeller, the youngest son of John D., Jr. David has been associated with Chase since 1946.[5]

Through Chase, through the family's investments, and through their trust funds, the Rockefellers have acquired interests in a host of companies. An example of the interaction of these various investment vehicles is the Rockefeller Center in New York city. The Center, built in the 1930s, was one of the largest investments made by the Rockefeller family. In 1974 it was valued at over $98 million. The family's investment in the Center is exceeded only by its investment in Exxon, yet

Chase Manhattan Bank is listed as the owner of the Center. The bank in fact owns 100 per cent of the Center's stock as a trustee for the family.[6] One can only speculate on the stock Chase Manhattan owns in a fiduciary capacity for the family in other companies.

A U.S. congressional committee in 1974 listed the securities held directly by members of the family and indirectly through two trust funds (established in 1934 and 1952 by John D., Jr. for the benefit of his children and grandchildren respectively).[7] It placed the total value of these assets at slightly over $1 billion,[8] and this list was by no means comprehensive.[9]

Among the interests not included in the congressional total were the holdings of the Rockefellers and the Chase Manhattan Bank in Eastern Airlines, in which Laurance Rockefeller, in particular, had substantial investments. The *New York Times* reported that "while the family had by no means a controlling interest in Eastern, a long association with the airline, coupled with Laurance Rockefeller's investment and that of the family bank, Chase-Manhattan, made it highly influential in determining the airline's policies."[10]

This *New York Times* story raised the important question of how much stock the Rockefellers do have to own to gain control of a company. The 1974 congressional data, for instance, show that the Rockefellers own only slightly more than one per cent of Exxon,[11] thereby rendering the impression that the family no longer controls the company created by John D. Rockefeller. Yet there is no question that the family still retains control of the company. As the *Washington Post* observed:

> The Rockefeller economic power may not exist in terms of a company-by-company analysis or in terms of outright control of particular companies. But it does exist in the minds of most Americans. Most businessmen would react quite differently if a Rockefeller-backed enterprise went into competition with them—or if they were invited into a joint participation with such in a venture—from the way they would if it were a John Doe-backed venture.[12]

In actuality, the shareholdings in Exxon of all the Rockefeller-controlled investment vehicles, including the Chase Manhattan Bank, constituted approximately 4 per cent of the company's outstanding stock in 1974, making the Rockefeller–Chase Manhattan group the single largest stockholder in the firm.[13] This bloc, while relatively small, is nevertheless sufficient to enable the Rockefeller–Chase group to exercise effective control over a company with such a widely-dispersed stock ownership as Exxon.[14]

### IBEC

Though there can be legitimate dispute about the extent to which the Rockefellers actually control many of the firms in which they have invested, there is no question about the control of one of them, International Basic Economy Corporation (IBEC). The Rockefellers owned about 79 per cent of the outstanding shares of that company.[15] Founded in Venezuela in 1947 for the ostensible purpose of promoting "the economic development of various parts of the world, to increase the production and availability of goods, things and services useful to the lives or lifelihood of their peoples, and thus to better their standard of living," it was initially financed by Creole, Shell, the Venezuelan government, and Nelson Rockefeller along with some members of his family. The two companies and the Venezuelan government received preferred shares while the Rockefellers received common stock, thereby acquiring direct control over the company.[16] The Venezuelan government and the oil companies later sold out their interests

**Table 11.4: Partial Listing of Financial Links of Rockefeller–Chase Manhattan Group to Major U.S. Companies, 1974**

***Oil***
Exxon[a]
ARCO[a]
Standard Oil of California
Mobil
Standard Oil (Indiana)
Marathon
Conoco
Texaco

***Machinery***
Caterpillar
Allis-Chalmers

***Pharmaceutical***
Merck
Johnson & Johnson
Upjohn
Warner-Lambert

***Retail Trade***
IBEC[a]
Sears-Roebuck
Kresge
Penney (J.C.)
Safeway

***Airlines***
United[a]
Northwestern[a]
National[a]
Overseas National
TWA
Delta
Braniff
Eastern

***Computers & Office Equipment***
IBM
Texas Instruments
Xerox
Sperry-Rand
Hewlett Packard

***Chemical & Photography***
Eastman Kodak
Dow
Allied Chemicals
DuPont
Monsanto
Hercules
Polaroid

***Other***
ALCOA
Archer-Daniels-Midland
AT&T
Coca-Cola
Corning Glass
Dun & Bradstreet
Long Island Lighting
Minnesota Mining & Manufacturing

***Electrical***
General Electric
Motorola
Reliance Electric
Westinghouse

***Media***
Columbia Broadcasting System
American Broadcasting Company
National Broadcasting Company
*New York Times*
*Time*
*Time-Life*
*Newsweek*

***Cosmetics***
Avon
Chesebrough-Pond's

***Forest & Paper Products***
International Paper
Weyerhauser

a. The group is the largest single shareholder.

Source: U.S. House of Representatives, Committee on the Judiciary, *Hearings, Nomination of Nelson A. Rockefeller*, 93rd Congress, 2nd Session (U.S. Government Printing Office, Washington, 1974), pp. 848–49; U.S. Senate, *Nomination of Nelson A. Rockefeller, Senate Executive Report No. 34* (U.S. Government Printing Office, Washington, 1974), pp. 569–79; and W. Greider & T. O'Toole, "Rockefeller Family Holdings Touch Every Economic Sphere," *Washington Post*, September 22, 1974.

after the company encountered severe financial problems.[17] Nelson Rockefeller reorganized the company and turned it into a giant multinational. He established a chain of supermarkets in Venezuela which, among other things, succeeded in driving many small Venezuelan retailers out of business. Under his leadership, IBEC acquired companies in many parts of the world, including the United States. By the end of the 1960s, it operated more than 140 subsidiaries in more than 33 countries. In addition to supermarkets, IBEC ran mutual funds, insurance companies, housing construction firms, and many others.[18] "IBEC," noted Collier and Horowitz, "was an atavar of a new business form—the U.S. multinational with subsidiaries and markets flung across the globe—which would be a primary fact in the economic life of the underdeveloped world in the second half of the twentieth century."[19] IBEC, they also noted, instead of building up the Venezuelan economy, "actually helped to make the country more dependent than before on U.S. corporations and the goods they offered."[20] IBEC thus turned out to be the vehicle through which the Rockefellers used their oil earnings to gain control over important sectors of the economies of Latin American countries.

### FINANCIAL LINKS

Rockefeller's oil wealth was also used to acquire control over a considerable portion of the U.S. economy, as the partial listing of the financial links of the Rockefeller–Chase Manhattan group, illustrated in Table 11.4, shows. This table, compiled from a number of sources, indicates that the group was the largest single shareholder in Exxon and ARCO, two supposedly competing oil companies, and in United, Northwestern, and National airlines, three other supposedly competing companies. The group's other holdings read like a "Who's Who" of American industry. It owns securities in at least eight oil companies and fifty-four non-oil companies. Though apparently absent from such important industries as steel, automobiles, and coal, the group has investments in leading companies in such key industries as computers and office equipment, chemicals, pharmaceuticals, electrical equipment, retail trade, and the news media. The group, furthermore, has financial links with a number of supposedly competing companies in most of these industries. To the extent that these financial links involve either outright control of these companies or extensive influence in them, the question must be raised as to how competitive these industries are. The group is certainly in a theoretical position to reduce competition among these companies by coordinating their activities.

**Table 11.5: Partial Listing of Directorship Links of Employees of Rockefeller Family and Associates, 1974**

| | |
|---|---|
| ***Transportation Equipment*** | ***Finance*** |
| American Motors | CIT Financial |
| Bendix | Greenwich Savings Bank |
| Chrysler | Lincoln First Bank |
| ***Airlines*** | ***Retail Trade*** |
| Eastern | IBEC |
| Seaboard World | Macy |
| | Kresge |
| ***Chemicals*** | ***Other*** |
| International Minerals & Chemicals | IBM |
| Mallindrodt Chemicals | Howard Johnson |
| | New York Telephone |
| | Rockefeller Center |

Source: U.S. House of Representatives, Committee on the Judiciary, *Hearings, Nomination of Nelson A. Rockefeller*, 93rd Congress, 2nd Session (U.S. Government Printing Office, Washington, 1974), pp. 769–70.

**Table 11.6: Corporations Represented on Board of Directors, Chase Manhattan Bank, 1974**

| *Automatic Controls* | *Other* |
|---|---|
| Hewlett Packard | Allied Chemicals |
| Honeywell | American Smelting & Refining |
| | AT&T |
| *Oil* | Burlington Industries |
| Exxon | Equitable Life Assurance |
| Royal Dutch Petroleum (Shell) | Federated Department Stores |
| Standard Oil (Indiana) | General Foods |

Source: W. Greider & T. O'Toole, "Rockefeller Holdings Touch Every Economic Sphere," *Washington Post*, September 22, 1974.

DIRECTORSHIP LINKS

One measure of the Rockefeller potential to influence these companies is membership on their boards. In their 1974 testimony before the congressional committee investigating the nomination of Nelson Rockefeller for the vice-presidency of the United States, Domhoff and Schwartz revealed that nine employees of Rockefeller Family and Associates sat on the boards of some forty corporations, "directing an aggregate of some 70 billion dollars in assets."[21] Their data, which do not include the interlocks between members of the Chase board and those of other companies, was, in their words, "undoubtedly incomplete."[22] Nevertheless, their data show that the Rockefeller influence extends to industries in which the group has no apparent equity interests, like transportation equipment. As can be seen in Table 11.5, Rockefeller representatives sat on the boards of several supposedly competing companies in such industries as transportation equipment, airlines, chemicals, finance, and retail trade. Again the question arises as to whether these representatives used their influence to coordinate the activities of the companies on whose boards they sat.

The Rockefellers were also in a position to influence other companies whose top officers sat on the board of Chase Manhattan. Among them were executives from three oil companies—Exxon, Royal Dutch Petroleum, and Standard Oil (Indiana)—and two companies in automatic equipment—Honeywell and Hewlett Packard. Those firms with officers on the Chase board are listed in Table 11.6. It should be noted that they include firms supposedly competing with companies shown in Tables 11.4 and 11.5.

*Non-Oil Energy*

These tables are incomplete in that they show only the linkages of the Rockefeller–Chase Manhattan group. They do not show the acquisitions by the firms controlled or influenced by this group, acquisitions which have significantly strengthened the economic power of the group over the years. A complete listing of all of them would be beyond the scope of this book. It, in fact, would require a book in itself. What is important, as far as this book is concerned, is a listing of the important acquisitions of some of the major oil companies, whether Rockefeller-controlled or not, as a means of showing how these companies have used their profits to extend their control over the economy.

The expansion of the oil companies started in the late 1950s and early 1960s with the acquisition of coal and uranium reserves. They obtained these reserves either through the purchase of companies owning coal or uranium properties or by buying these properties outright. By the early 1970s this trend had become so pronounced in the United States that it caused concern among the agencies enforcing the antitrust laws. A study prepared by the Federal Trade Commission (FTC) expressed these fears in the following words:

> Some observers have expressed concern over this trend since it has coincided with an increased flexibility of fuel use which has placed all four fuels [crude oil, natural gas, coal, and uranium] in competition with each other in a large number of energy consumption sectors. Establishment of the energy companies [companies exploiting more than one energy resource] has thus raised the issue of whether this movement is serving to stifle competition by reducing the number of independent decision making units that operate within the energy sector.[23]

By the time the FTC conducted its study, five oil companies had already joined the ranks of the twenty companies holding the largest coal reserves in the United States. These five companies, as is shown in Table 11.7, controlled almost one-third of the reserves owned by the top twenty. The Rockefeller–Chase Manhattan group had shares in at least two of these companies, Conoco and Exxon, ranked fourth and fifth among firms holding coal reserves.

A similar situation existed in uranium. In 1971, seven companies controlled 70 per cent of U.S. uranium reserves. Among these seven were four oil companies—Getty, Gulf, Exxon, and Kerr-McGee.[24] Of the fifteen top U.S. oil companies, ten had both coal and uranium facilities in the early 1970s; two had only coal facilities; and three had only uranium facilities. Of these fifteen companies, the Rockefeller–Chase

**Table 11.7: Oil Companies' Share of Coal Reserves Owned by 20 Largest Coal Companies in the United States, 1970**

| *Company* | *Reserves* (Billion Tons) | *Rank* |
|---|---|---|
| Conoco[a] | 7.7 | 4 |
| Exxon[b] | 7.0 | 5 |
| Occidental | 3.3 | 7 |
| Gulf | 2.6 | 9 |
| Kerr-McGee | 1.5 | 15 |
| Total | 22.1 | |
| Total Reserves of Top 20 | 74.7 | |
| Per Cent Held by Oil Companies | 29.6 | |

a. Rockefeller–Chase Manhattan group has shares in this company.
b. Rockefeller–Chase Manhattan group largest single shareholder in this company. Company represented on board of Chase Manhattan Bank.

Source: U.S. Federal Trade Commission, *Concentration Levels and Trends in the Energy Sector of the U.S. Economy* (Federal Trade Commission, Washington, 1974), p. 213.

Manhattan group was a stockholder in seven and the largest single shareholder in two, Exxon and ARCO. Three of these companies—Exxon, Shell, and Standard Oil (Indiana)—had representatives on Chase's board. The companies and their principal subsidiaries in coal or uranium mining are listed in Table 11.8.

Though the oil companies had made significant inroads in coal and uranium by the early 1970s, the FTC report concluded that concentration in the latter two industries was still relatively low. Nevertheless it warned:

> If petroleum companies are acquiring significant amounts of coal and uranium reserves, then in the future they may control a rising fraction of coal and uranium production. In that case, energy production concentration may rise significantly as petroleum companies become more dominant in coal and uranium production.[25]

This increased concentration, the report pointed out, could lead to "reduced interfuel competition with a consequent rise in the overall price of fuels." Furthermore, it could retard technological advance since the "petroleum firms have no incentive to pursue those technological advances in coal and uranium which represent a threat to their oil and gas operations."[26]

**Table 11.8: Partial List of Major Oil Companies Owning Coal and Uranium Facilities, 1970 (Coal) and 1972 (Uranium)**

| *Company* | *Subsidiaries* |
|---|---|
| | ***Both Coal and Uranium*** |
| Exxon[a] | Monterey Coal Co.<br>Jersey Nuclear |
| Texaco[b] | Texas-Zinc (50%) |
| Gulf | Pittsburgh & Midway Coal<br>C&K Coal |
| Mobil[b] | -------- |
| ARCO[c] | -------- |
| Getty | KGS-JV (33%)<br>Petrotomics (33%) |
| Sun | -------- |
| Conoco[b] | Consolidation Coal |
| Kerr-McGee | Petrotomics (50%)<br>KGS-JV (50%)<br>Kermac Nuclear Fuels<br>Ambrosia Lake Uranium<br>Spencer Chemical<br>American Lake Uranium<br>Guyan Eagle<br>Lakeview Mines |
| Sohio | Old Ben Coal |
| | ***Coal Only*** |
| Shell[d] | -------- |
| Standard Oil of California[b] | -------- |
| | ***Uranium Only*** |
| Standard Oil (Indiana)[e] | -------- |
| Phillips | Holly Minerals |
| Cities Service | -------- |

a. Rockefeller–Chase Manhattan group is largest single shareholder. Represented on board of Chase Manhattan.
b. Rockefeller–Chase Manhattan group is a shareholder in company.
c. Rockefeller–Chase Manhattan group is largest shareholder.
d. Represented on board of Chase Manhattan.
e. Rockefeller–Chase Manhattan group is a shareholder. Represented on board of Chase Manhattan.

Sources: U.S. Federal Trade Commission, *Concentration Levels and Trends in the Energy Sector of the U.S. Economy* (Federal Trade Commission, Washington, 1974), pp. 175, 208, 211, 213, 217–18, 233, and M. Tanzer, *The Race for Resources, Continuing Struggles over Minerals and Fuels* (Monthly Review Press, New York and London, 1980), pp. 116–17.

Since the report was written, the oil companies, with their treasuries bulging with increased cash flows stemming from higher energy prices, have continued their penetration into coal and uranium. They have also gone into other energy fields, like solar energy. Some, like Gulf, have dropped the word *Oil* from their names. Gulf is now officially called the Gulf Company rather than the Gulf Oil Company, indicating that it now considers itself an energy rather than an oil company.

*Non-Renewable Resources*

Some companies have expanded from energy into the entire non-renewable resource field. Anticipating that the prices of these resources will rise as they become relatively scarce over time, the companies are looking to them as sources of future profits.

The 1970s witnessed a number of purchases by oil companies of non-fuel mineral companies. Standard Oil (Indiana), for instance, entered into a nickel mining venture in Australia in 1973.[27] In the same year it joined a copper mining venture in Canada with Cominco.[28] Standard Oil of California purchased a one-third interest in a nickel ore project in Colombia in 1974.[29] In 1975, for $333 million, it bought 20 per cent of the common stock of AMAX, one of the world's largest mining companies,[30] a major producer of molybdenum, iron ore, coal, nickel, copper, zinc, and potash as well as one renewable resource, forestry products.[31] Apparently not satisfied with its 20 per cent share, Standard offered to buy out all the shares for the sum of $3.89 billion in early 1981. The other stockholders rejected this take-over attempt.[32]

In 1977 Standard Oil of California joined Amoco Mineral, a subsidiary of Standard Oil (Indiana) in forming a joint venture, Pacific Nickel Indonesia, to explore for and exploit nickel resources in Indonesia.[33] Another company interested in non-energy minerals is Gulf. In 1973 it established Gulf Energy and Minerals, a subsidiary to develop both non-petroleum energy resources and non-energy mineral resources.[34] In 1978 it bought a 51 per cent interest in Silver King Mines' Ward Mountain district.[35]

Another company entering the non-energy minerals industry is Sohio. In 1981 Sohio paid $1.77 billion to acquire Kennecott, the leading U.S. copper producer.[36] Kennecott, through its ownership of Peabody Coal, was the third largest owner of coal reserves in the United States in 1970.[37] This purchase thus gave Sohio extensive coal as well as copper and other mineral reserves.

In 1980 Sohio's parent, British Petroleum, made the largest financial transaction in London's history when it bought the huge British mining company, Selection Trust, for more than $1 billion. British Petroleum owns 53 per cent of Sohio's shares.[38] The British Petroleum-Sohio complex hence has access to a significant store of the world's mineral resources.

Exxon has owned its own mineral-bearing properties for many years. In 1976 it found copper and zinc on its lands in Wisconsin. This find could make it one of the top ten mineral producers in the United States.[39] Its Canadian subsidiary, Imperial Oil, has acquired a lead and zinc mine in Nova Scotia and a copper mine in British Columbia. It is also taking part in joint venture drilling for molybdenum in British Columbia.[40] Imperial's metal activities have become so important that Imperial has created a special division, Esso Minerals Canada, to handle them.[41] Its parent, Exxon, created a subsidiary, Exxon Minerals USA, in 1977 for the same purpose.[42] In 1980 it consolidated all its mineral activities into a new unit, Exxon Minerals Group.[43]

Exxon's acquisitions, however, have not been nearly as important as those of ARCO, the other oil company in which the Rockefeller–Chase Manhattan group is the largest single shareholder. In 1977 ARCO bought out Anaconda, the ninth largest copper producer in the world,[44] for $536 million in cash.[45] The combined assets of the merged company were estimated to be $9.4 billion.[46]

*General Diversification*

ARCO's expansion, moreover, was not confined to the minerals industries. In 1976 it surprised financial circles by buying a prestigious London newspaper, the *Observer*.[47] The acquisition of this paper had the potential of significantly expanding the company's influence. The media, after all, can have an important impact on public opinion and government policy. Apparently ARCO found that the *Observer* was too expensive a vehicle to use for this purpose; it did not turn out to be a profitable venture, and ARCO sold it in 1981.

Besides newspapers, ARCO also entered transportation. In 1978, it bought a one-third interest in R. W. Miller, an Australian coal firm with shipping interests.[48] In 1979 it acquired all the shares of Northrup, a heating and air conditioning company which also produces solar energy equipment.[49]

ARCO was not the only oil company to stray far afield. It was emulated by Gulf, among others. In 1974 Gulf established a special unit, Gulf Merchandising, to compete in the auto after-market.[50] In 1976 it formed a joint venture, American Heavy Lift, with Hansa Lines to build and operate ships. Gulf owned 75 per cent of the joint company.[51] Gulf, however, was not successful in all its attempts to buy out other firms. After attempting to purchase CNA Financial in 1973, it indefinitely postponed merger talks with that company.[52] It also failed in its endeavours to acquire Rockwell, a machinery manufacturer, in 1975.[53] Perhaps most interesting of all was its lack of success in trying to buy Ringling Brothers' Circus from Mattel, the toy manufacturer, in 1974.[54] Apparently Gulf had considerable difficulty in finding suitable investment outlets for its bulging profits.

Exxon was another company which moved into other industries. In 1979 it purchased Xerox's office products division.[55] Late in 1979 it acquired control of Reliance Electric, an electrical machinery manufacturer in which the Rockefeller–Chase Manhattan group has an interest, for $1.76 billion. Before Exxon was able to complete this purchase it had to overcome the opposition of the FTC, which finally gave its consent to the merger after a long period of negotiations.[56] As Exxon expanded into new ventures, it set up a special unit, Exxon Enterprises, to handle its non-mineral activities. In 1978 Exxon Enterprises established Data Screen to manufacture multiplex LCDs.[57] In 1979 it created Optical Information Systems to produce semiconductors and laser components.[58] One year later it formed an office system subsidiary, Exxon Information Systems.[59]

Exxon's successes in moving into new territories have not been as notable as those of Mobil. Mobil has become the most diversified of all the major oil companies through the purchases of W. F. Hall Printing Company for $50.5 million in 1978[60] and, earlier, of Marcor, a holding company whose principal subsidiaries were Montgomery-Ward, a major U.S. retailer, and Container Corporation of America, a large packaging firm. The Marcor acquisition cost approximately $800 million in 1976.[61] As with Exxon's purchase of Reliance, the U.S. antitrust authorities initially opposed Mobil's attempts to buy Marcor. After protracted negotiations, they finally gave up in their efforts to block the merger.[62] Mobil has thus become a major firm in U.S. retailing and packaging.

What we have witnessed over the past few decades is a metamorphosis of the oil companies. During the late 1950s and the 1960s they gradually began to transform themselves into energy companies through the acquisition of coal and uranium facilities. Following the rise in oil prices in the 1970s, they first became mineral companies through their purchases of properties containing copper, lead, zinc, nickel, and other non-ferrous metals. They later became true conglomerates through their expansion into manufacturing.

**Table 11.9: Entry of Some Major Oil Companies into Fields Other than Oil and Petrochemicals since 1970 (Partial Listing)**

**ARCO**[a]

NON-OIL ENERGY

Forms Cenar Association, joint venture in uranium.

Joins Coalcon, joint venture to convert coal to gas and liquids.

Invests 50% in Centennial Hydrocarbons, joint venture with DuPont.[b]

Buys Solar Technology International.

Buys oil shale holdings of Equity Oil.

Joins Colony Development, oil shale firm.[c]

Buys Utah coal operations and reserves of General Exploration.

Buys interest in R. W. Miller, a major Australian coal concern.

Buys Swisher Coal, renamed Beaver Coal.

NON-ENERGY MINERAL

Purchases Anaconda Copper.

OTHER

Buys *Observer*, London newspaper.[d]

Buys Northrup, air conditioning and heating firm.

Buys Wisconsin Centrifugal.

Acquires R. N. Parsons & Sons, Parsons Transport, Parsons Leasing, and Parsons Terminal.

**Exxon**[e]

NON-OIL ENERGY

Forms joint uranium enrichment venture with ARCO.

Joins Anvil Points shale oil extraction project.

Joins Avco in joint uranium project.

Forms Jersey Nuclear-Avco enterprise.

Exxon Nuclear joins GE in enriched uranium venture. GE later drops out.

With Esso Standard Sekiyu, Japan, creates Japan Coal Liquefaction Company to build plant in Texas.

Buys ARCO's 60% share in Colony Development, oil shale firm.

Buys 25% in Hail Creek Project, Australia.

Acquires Yeelirri Uranium Deposit, Australia.

Forms Société des Applications de l'Hélioénergie with Thomas CSF of Paris, France.

Invests 50% in joint venture to develop uranium deposits in Midwest Lake, Saskatchewan, through its Canadian subsidiary, Imperial Oil.

Imperial Oil in joint venture to develop coal in British Columbia.

Imperial Oil has interests in heavy oil development, Cold Lake, Alberta.

Imperial Oil partner in Syncrude, joint venture to produce oil from tar sands, Alberta.

Imperial Oil purchases Byron Creek Collieries, a British Columbia coal mining company.

NON-ENERGY MINERAL

Imperial Oil acquires Granduc copper mine in British Columbia.

Imperial Oil buys lead-zinc mine in Gays River, Nova Scotia.

Imperial Oil in joint venture to develop molybdenum in Trout Lake, British Columbia.

Imperial Oil in joint venture with Japanese Sumitomo interests to develop copper, zinc, lead, and silver deposits in Kutcho Creek, British Columbia.

Exxon buys 87% of Columbia Minera Disputada de las Condes, a Chilean copper company.

OTHER

Buys Xerox's office products division.

Exxon Enterprises forms firm, Data Screen, to manufacture multiplex LCDS.

Exxon Enterprises forms Optical Information Systems, a semiconductor, laser component firm.

Exxon Enterprises forms Exxon Information Services, an office information subsidiary.

### Gulf

NON-OIL ENERGY

Forms Gulf Energy and Minerals, a non-oil energy subsidiary.

Acquires Gulf United Nuclear.

Buys United Nuclear Fuels.

Forms Combustible pour Reacteurs a Haute, to manufacture nuclear fuels in France.

Forms General Atomic with Shell.

Forms Solvent Refined Coal International with Ruhrkohle, a German firm, and Mitsui, a Japanese firm, to build coal liquefaction plant.

NON-ENERGY MINERAL

Buys 51% share in Silver King Mine's Ward Mountain mining district.

Gulf Canada, Canadian subsidiary, acquires coal interests in Belcourt, British Columbia, and Chip Lake, Alberta.

Obtains 25% partnership interest in Oklahoma Nitrogen Co.

OTHER

Subsidiary merges with Major Realty.

Forms Gulf Merchandising to compete in auto after-market.

Forms American Heavy lift with Hansa to build and operate ships. Gulf holds 75% interest.

Purchases Kewanee Industries.

### Mobil[f]

NON-OIL ENERGY

Forms company with Tyco Labs to develop silicon ribbon technology for solar cells. Mobil's investment is 80%.

Joins joint venture, Coalcon, to develop $237-million coal company.

Buys Mt. Olive & Staunton's coal assets in Illinois for $47.5 million.

OTHER

Buys Marcor for approximately $800 million. Marcor's chief subsidiaries are Montgomery-Ward, a major retailer, and Container Corporation of America, a large packaging company.

Buys W. F. Hall Printing for $50.5 million.

### Phillips

NON-OIL ENERGY

Organizes a subsidiary, Phillips Coal.

Joins with Sun and Sohio in a joint venture to develop shale oil in Utah.

Forms joint venture, Acurex Solar Energy, to develop solar energy systems.

Forms Phillips Uranium.

Agrees to build geothermal power plant for Utah Power and Light.

### Shell-Royal Dutch Shell Group[g]

NON-OIL ENERGY

Forms General Atomic with Gulf.

Buys Seaway Coal for $65 million.

Buys 50% of Massey Coal for $680 million.

Buys Crow's Nest Coal in British Columbia.

OTHER

Buys 50.5% of Alpha Text, Canadian computerized information services concern.

### Sohio-BP Group

NON-OIL ENERGY

Joins with Phillips and Sun in a joint venture to develop shale oil in Utah.

Engaged in uranium exploration in Canada.

Purchases coal properties in British Columbia.

NON-ENERGY MINERAL

Purchases Kennecott Copper for $1.77 billion.

Purchases Selection Trust for $1 billion.

### Standard Oil of California[f]

NON-OIL ENERGY

Forms joint venture with Geothermal Resources.

Buys American Nuclear for $30 million.

NON-ENERGY MINERAL

Acquires one-third interest in nickel ore project in Colombia.

Buys 20% common stock of AMAX, major mineral producer, for $333 million.

Forms Pacific Nickel Indonesia with Amoco, subsidiary of Standard Oil (Indiana).

### Standard Oil (Indiana)[h]

NON-OIL ENERGY

Forms Rio Blanco Oil Shale as a joint venture.

Buys 21% of Solarex, manufacturer of solar energy systems.

NON-ENERGY MINERAL

Joins nickel venture in Australia.

Joins Caribou copper mining venture with Cominco in Canada.

Forms Pacific Nickel Indonesia with Standard Oil of California

Has gold reserves in Detour Lake area, Ontario.

Buys Cyprus Mines.

OTHER

Jointly forms Analog Devices/Enterprises to lend expansion capital.

### Sun

NON-OIL ENERGY

Joins with Phillips and Sohio in a joint venture to develop shale oil in Utah.

Forms subsidiary, Sunoco Energy Development, for synthetic fuels.

OTHER

Buys Stop-N-Go Foods.

Buys Data Transmission for $30 million.

Acquires SJT, St. Johnsbury Trucking Co., through joint venture.

Acquires HP International, industrial distribution firm.

Acquires Carolina Co. for cash.

Acquires Kar Products.

Acquires Belton Dickinson, pharmaceutical concern.

Acquires Atlas Screw and Specialty.

Buys Weiland Computer Group.

Buys Mr. Zip.

### Texaco[f]

NON-OIL ENERGY

Buys Wyoming coal lands of Reynolds Metals.

OTHER

Buys General Automatic Oil Heating.

---

a. Rockefeller–Chase Manhattan group single largest stockholder in company.
b. ARCO pulled out of this joint venture in 1977.
c. Exxon purchased ARCO's 60 per cent share of Colony Development in 1980.
d. ARCO sold the *Observer* in 1981.
e. Rockefeller–Chase Manhattan group largest single stockholder in company. Company represented on board of Chase Manhattan.

Table 11.9 gives an indication of the scope of this expansion. It lists the acquisitions of some of the major companies in industries other than oil and petrochemicals, the traditional hunting grounds of the oil companies, since 1970. This list is by no means complete in that it does not contain all the acquisitions by all the companies; it must be viewed merely as the tip of the iceberg.

The significance of this list lies in the implications of the investment provisions of the agreement establishing the IEA. Given the guarantees of a floor price and investment incentives for the energy companies, one can expect that these companies will assume a greater and greater control over the economies of the member countries as long as the investment provisions are observed. These provisions, which effectively ensure the profitability of these companies, will afford them the opportunity to acquire more and more companies in a wide array of industries throughout the non-Communist world. The profits from these industries will further swell the flow of funds accruing to the oil companies, thereby enhancing their ability to acquire more firms. An example of this process is the reported after-tax profits of the U.S. oil companies in 1980. In that year, eight of the top ten profit-makers in the U.S. were oil companies. In the previous year only five oil companies were in the top ten.[63] A continuation of this shift of profits toward the oil companies will increase to a marked degree the economic power of these companies and of the people who control them. The power of the Rockefeller–Chase Manhattan group, in particular, will grow significantly. With this increase in economic power we can expect an increase in political power.

## Political Links

Within the United States the political power of the oil companies grew as their economic power increased. No longer were the Rockefellers the pariahs of the Establishment, as they had been in 1911 when the U.S. Supreme Court upheld the order dissolving the old Standard Oil Company. The Rockefellers have become the Establishment. The oil companies, in general, and the Rockefellers, in particular, have increasingly placed their imprint on all facets of American life.

---

f. Rockefeller–Chase Manhattan group owns shares in company.

g. Company represented on board of Chase Manhattan.

h. Rockefeller–Chase Manhattan group owns shares in company. Company represented on board of Chase Manhattan.

Sources: Company reports, newspapers, and trade journals.

**Table 11.10: Links between U.S. Secretaries of State and Rockefeller–Chase Manhattan Group, Post–World War II Years**

| *Secretary* | *Link* |
|---|---|
| Dean Acheson | Former partner, Covington and Burling, a law firm of Standard Oil (New Jersey). |
| John Foster Dulles | Former partner, Sullivan and Cromwell, major law firm of Standard Oil (New Jersey). Chairman, Board of Trustees, Rockefeller Foundation. |
| Dean Rusk | President, Rockefeller Foundation. |
| Henry Kissinger | Chairman, International Advisory Committee, Chase Manhattan Bank. Director, Special Studies Project, Rockefeller Brothers Fund. Received $50,000 gift from Nelson Rockefeller before going to Washington to assume governmental duties. |
| Cyrus Vance | Chairman, Board of Trustees, Rockefeller Foundation. |
| Alexander Haig | Brought to Washington in 1969 by Kissinger to serve as Kissinger's assistant. |

Source: *Who's Who in America*, various editions, and *New York Times*, October 6, 1964 and December 17, 1980.

Probably they were nowhere more successful than in the field of U.S. foreign policy. This book has already shown how the U.S. government after World War I pursued a policy of backing the U.S. oil companies in their overseas ventures. It has also shown how, in Saudi Arabia, U.S. State Department officials tended to favour the major U.S. oil companies, like ARAMCO, over the independents. A similar pattern was exhibited in Iran. As a result, very close links have developed between the State Department and the major U.S. oil companies in the post–World War II era.

The links between the department and the Rockefeller interests were especially close. As can be seen in Table 11.10, six secretaries of state had some connections with the Rockefellers. These secretaries served under each of the post–World War II presidents: Acheson under Truman, Dulles under Eisenhower, Rusk under Kennedy and Johnson, Kissinger under Nixon and Ford, Vance under Carter, and Haig under Reagan. These links have thus remained unbroken under both Republican and Democratic presidents. Such a continuous link in a democracy, which ostensibly eschews heredity as a basis for power, is indeed remarkable. Probably no other single family in modern history has had such an influence on the foreign policy of a major world power.

The links to the oil companies furthermore extended beyond the

Rockefeller group. In 1982 President Reagan, following the resignation of Alexander Haig, appointed George Shultz as secretary of state. Shultz was president of Bechtel Corporation, which constructs refineries and pipelines for oil companies throughout the world. As mentioned earlier, it was Bechtel that, in partnership with Standard Oil (New Jersey), constructed the Trans Mountain pipeline from Edmonton to the West Coast. Another important figure with Bechtel connections was John J. McCone. A salesman for the company during World War II, he later held such posts as under-secretary of defense, chairman of the Atomic Energy Commission, and director of the CIA.

The relations between Bechtel and the Reagan administration have been especially close. In addition to Shultz, Secretary of Defense Caspar Weinberger and Deputy Secretary of Energy W. Kenneth Davis were former Bechtel employees. Philip Habib, Reagan's special envoy to the Middle East, joined Bechtel after retiring from the State Department. While serving as special envoy, he retained his position as a consultant to the company.

Other links between the oil companies and the State Department are shown in Table 11.11, which lists only some of the officials who participated in making foreign policy decisions. One of these was Zbigniew Brzezinski, President Carter's foreign policy adviser. Brzezinski was a director of the Trilateral Commission, a "think tank" founded by David Rockefeller.[64] According to its official statement, the commission "was formed in 1973 by private citizens of Western Europe, Japan and North America to foster closer cooperation among these three regions on common problems."[65] These "private citizens," it should be noted, represent powerful sections of the business elites of their respective countries. It was certainly to the interests of the Rockefellers to gain their acquiescence in matters of vital concern to the family. Brzezinski presumably served this purpose when he was a director of the commission; one can only speculate as to whether he continued to serve this purpose when he became an adviser to the president of the U.S.

Of those mentioned in Table 11.11, Levy is probably the best example of the close links between the oil companies and the State Department. Levy, whom the *New York Times* dubbed "the dean of oil consultants,"[66] was, as already noted, the oil director of the Marshall Plan. It was under his administration that Marshall Plan funds were used to find markets for American-owned oil in Europe. He was also adviser to W. Averill Harriman in Iran during the oil crisis there in the early 1950s. Apparently the Shah of Iran greatly appreciated Levy's advice, and in 1969 the Shah awarded him a high decoration.[67]

**Table 11.11: Links of Some Key U.S. Foreign Policy Resource Personnel to Oil Interests, Post–World War II Years**

| *Person* | *Link* |
| --- | --- |
| *Winthrop Aldrich*, U.S. ambassador to Britain. | Brother-in-law of John D. Rockefeller, Jr.; head, Chase Manhattan Bank. |
| *Zbigniew Brzezinski*, President Carter's Assistant for National Security Affairs. | Director of Trilateral Commission. |
| *George Bush*, U.S. vice-president under President Reagan and adviser to Reagan on foreign policy; U.S. Ambassador to UN; chief U.S. liaison officer to China; director, Central Intelligence Agency. | Co-founder and director, Zapata Petroleum Co.; president and chairman, Zapata Offshore Co.; member, Trilateral Commission. |
| *Emil Collado*, various State Department positions, including position in Office for Fuel Development; U.S. executive director, World Bank; trustee, Export-Import Bank. | Executive vice-president of Exxon. (Joined Exxon after leaving State Department.) |
| *Arthur Dean*, U.S. Delegate to UN; chairman, U.S. delegation to Law of the Sea Conference, Geneva, 1958–60; special U.S. ambassador to Korea; chairman, U.S. delegation to Disarmament Conference, 1962. | Partner, Sullivan and Cromwell, main law firm of Standard Oil (New Jersey); director, El Paso Natural Gas. |

In his capacity as a private consultant, Levy had as his clients the world's major oil companies and various governments, including the governments of Venezuela, Canada, Alaska, Alberta, Iran, India, and the United States. He has also advised such international agencies as the World Bank, the United Nations, and the European Economic Community.[68]

The advice he rendered to some of these governments may have been of questionable value. For instance, it was largely on Levy's advice that Canada adopted the NOP.[69] Levy also advised India to refrain from exploring for oil, instead letting the international oil companies perform this function.[70] This was the general advice Levy gave to all underdevel-

| | |
|---|---|
| *Allen Dulles*,<br>director, Central Intelligence Agency. | Chief counsel, Sullivan and Cromwell. |
| *Herbert Hoover, Jr.*,<br>under-secretary of state; chief architect of Iranian consortium; special petroleum adviser to John Foster Dulles. | Petroleum engineer; director, Union Oil Company. |
| *Walter J. Levy*,<br>oil director, Economic Cooperation Administration; consultant to various governments; consultant to Harriman on Iranian mission; consultant to World Bank, UN, and European Economic Community; consultant to U.S. State Department; decorated by Shah of Iran; decorated by Secretary of State Dean Rusk. | Leading oil industry consultant; has most major oil firms as clients. |
| *John J. McCloy*,<br>president, World Bank; U.S. military governor and high commissioner, Germany; adviser to President Kennedy. | Chairman of board, Chase Manhattan Bank; member law firm, Milbank, Tweed, Hadley and McCloy. (Firm has defended all the major international oil companies against antitrust actions.) |
| *George McGhee*,<br>counsellor to the National Security Council; various State Department posts; senior adviser, North Atlantic Treaty Council; U.S. ambassador to Turkey; U.S. ambassador to Germany, U.S. ambassador-at-large. | Sole owner, McGhee Petroleum Co.; director, Mobil Oil. |

Sources: *Who's Who in America* and *Who Was Who in America*, various issues, and newspaper and magazine clippings.

oped countries, and it was most explicitly stated in his famous study for the World Bank, *The Search for Oil in Developing Countries*. His whole theme was that developing countries should not devote their scarce resources to the highly risky undertaking of exploration; the international oil companies were better equipped to assume this risk; the developing countries should instead concentrate on improving health, education, and the quality of life. The acceptance of Levy's advice would thus give the international oil companies control of the developing nations' oil resources.

His advice was effectively refuted by Tanzer, who had worked for

Esso Standard Eastern in India. Tanzer wrote:

> The question . . . of whether a government should "go it alone" cannot be solved *a priori* but only in the light of specific possibilities and resources facing the underdeveloped country. There should be no presumption that oil exploration is beyond the means of a country nor that if the government does not undertake it private companies will. To the contrary, the presumption should be that an underdeveloped country cannot afford not to undertake an oil exploration program. This is because, viewed in the light of a social return on investment analysis, oil exploration has uniquely attractive features for economic development.[71]

Tanzer also pointed out that the question of risk has been greatly exaggerated because "the risk can be reduced considerably by relatively inexpensive preliminary surveying efforts."[72] He noted that

> exploration is not an all-or-nothing thing, but a series of steps seeking information, which can be cut short if the initial information indicates that prospects are dim. Thus, if four wells cost $10 million, but the first two give very bad results, then you can cut your losses by ending exploration, and hence your actual risk capital will turn out to be much less than the maximum.[73]

It would thus seem that the advice that Levy gave so willingly to India and other Third World countries would prevent them from using oil as an instrument of economic development. By overexaggerating the "risks" of exploration, he urged them to adopt a policy that would simultaneously ensure that (1) the international oil companies would gain control of their oil resources and (2) the countries would remain in a perpetual state of underdevelopment. In the event, India ignored Levy's advice and set up a national oil company, the Oil and Natural Gas Commission, which did find oil.[74]

Levy also gave advice to the United States government, for which he received no compensation. In 1968, he served as an unpaid consultant to: Eugene Rostow, under-secretary of state for political affairs; Anthony Solomon, assistant secretary of state for economic affairs; Covey Oliver, assistant secretary of state for inter-American affairs; and Dean Rusk, secretary of state.[75] Secretary of State Dean Rusk in 1968 presented him with a plaque in recognition of his twenty-five years of service to the State Department. Its words read, "In grateful appreciation for your invaluable contribution to the welfare of the United States."[76] Dean Rusk was not the only secretary of state who placed a high value on Levy's services. Turner claims that Levy was the principal author of a speech Kissinger gave in December, 1973 advocating the creation of an international energy agency.[77]

Perhaps most important of all, Levy's services have been highly valued by the international oil companies. *Business Week* quoted one unnamed top executive of a major international oil company as praising Levy for being "good, honest, and forthright."[78] This is unquestionably an accurate description as far as the oil companies and the United States government are concerned.

Levy's activities, along with those of the others listed in Table 11.11, illustrate how over the years U.S. foreign policy has developed into an oil foreign policy, a policy designed to protect American oil interests throughout the world. As Levy put it in discussing the Middle East, "What is at stake is not European oil supplies, not American oil supplies, but American ownership of oil."[79]

Nationalizations have put an end to American ownership both in the Middle East and in other oil-producing areas. But these nationalizations need not prove disastrous to U.S. interests provided that the United States can gain control of non-OPEC energy sources. The achievement of this control will enable it to re-establish its hegemony in energy.

If this hegemony is re-established and OPEC prices fall, then the IEA "floor price" provisions may be invoked. But the financial interests are not satisfied with the IEA's present "floor price" of $7 a barrel. They want a much higher guaranteed price. In addition they want government financial aid for private energy projects. A 1978 report by the Trilateral Commission, for instance, states:

> If prices remain at or near current levels, there will be little economic incentive to develop readily substitutable alternative sources or to reduce demand growth as rapidly as would be desirable under the more pessimistic longer-term forecasts. For this reason, it may be appropriate for the Trilateral countries to (1) consider various mechanisms for increasing prices gradually over the next several years in those areas of the economy where it is desirable to encourage conservation and (2) provide special incentives to industry (e.g. loan guarantees, price guarantees, etc.) for the development of alternative sources of energy.[80]

There is little doubt that, given the close ties between the U.S. oil companies and the State Department, the U.S. government will exert considerable pressure on other consuming nations to adopt measures similar to those suggested by the Trilateral Commission. "You could never implement the energy policy as a purely economic matter," Kissinger noted. "It has," he continued, "been a foreign policy matter from the beginning."[81]

## The Era of Economic Warfare

In its drive to regain dominance, the U.S. has entered an era of economic warfare. This warfare is directed against the OPEC countries, the non-

OPEC oil-producing nations, and the rest of the industrialized world. This struggle, whatever its outcome, will be a long and costly one to all sides.

### *The Drive against* OPEC

One of the principal U.S. aims is to destroy or seriously weaken OPEC. The OPEC members, these Schumpeterian imitators, are the new entrants to world capitalism, threatening to topple the established entrepreneurs from the pinnacles of power. The U.S. is determined to smash them.

It intends to do this by re-creating a surplus of energy under U.S. control. This surplus can be achieved through the development of non-OPEC energy sources or through a reduction in demand for OPEC energy or through some combination of both. Kissinger has stated that "the most important" aim of U.S. energy policy is "to bring in alternative sources of energy as rapidly as possible so that the combination of new discoveries of oil, new oil-producing countries, and new sources of energy create a supply situation in which it will be increasingly difficult for the cartel to operate."[82]

The introduction of new energy sources would bring downward pressure on oil prices. OPEC's only defence would be to cut back production. But, Kissinger contended, OPEC is limited in its ability to reduce production significantly. Kissinger reasoned as follows:

> Many producers are dependent on their revenues for economic development. Countries that can cut production painlessly are those that are simply piling up balances. Countries that need oil revenues for their economic development, like Algeria, Iran, and Venezuela, do not have an unlimited capacity to cut their production. If the production of these countries is cut by any significant percentage, their whole economic development plan will be in severe jeopardy. Therefore, the problem of distributing the cuts is going to become more and more severe.[83]

The severity of "the problem of distributing the cuts" might possibly lead to the break-up of OPEC as individual producing states slash prices to sell the extra barrel of oil. This could be the result of the nationalizations and the consequent elimination of the production-coordinating function performed by the international oil companies.

### *Non-*OPEC *Oil Countries*

The most immediate prospects for alternative energy sources lie in the non-OPEC oil countries—Great Britain, Canada, Norway, and Mexico. All but the last two are full members of the IEA. Norway, as mentioned earlier, is only an associate member. Mexico has not joined the IEA.

Given the rules of the IEA, Britain and Canada could have their energy resources taken over by U.S. companies. The rules of the IEA conflict with attempts of member countries to give priority to their own national companies, British National Oil Corporation in Britain and Petro-Canada in Canada. Norway's Statoil and Mexico's Pemex (Petroleos Mexicanos) also stand as barriers to U.S. penetration of their countries. U.S. policy-makers do not want these countries to develop their own national energy policies. Canada's attempts to change its relationship with the U.S. in the matter of oil policy have been detailed in the previous chapter. The other non-OPEC oil countries are also attempting to prevent the U.S. from controlling their oil, and the U.S. likes their independent attitude no more than it did Canada's.

### MEXICO

Because Mexico nationalized its oil in 1938, it does not have the problem of foreign ownership facing Canada. Nevertheless, the United States became interested in Mexican oil after the large discoveries made by Pemex in the mid-1970s. Though nationalized, Mexican oil was an alternative to OPEC oil and was much nearer to the United States than oil from the OPEC countries. It had the potential of being used by the U.S. in its bargaining with OPEC.

Despite efforts by the U.S. and other industrialized countries to make Mexico increase its production, the Mexican government has adopted a deliberate policy of limiting its output to between 2.2 and 2.5 million barrels per day.[84] It has also limited its exports to about 1.5 million barrels per day.[85] Mexico furthermore forced the U.S. to buy its natural gas at a relatively high price as a condition for buying more oil.[86] Mexico has demanded that the U.S. give its manufactured exports special treatment in the U.S. market and that it accept more Mexican immigrants. The U.S. is insisting that Mexico abide by the rules of the General Agreement on Trade and Tariffs (GATT), which prohibits such special treatment. Mexico has refused to join GATT.[87]

There is no question that the U.S. regards Mexican oil as a very important part of its overall energy strategy. It is no coincidence that President Reagan arranged to meet Mexico's President Lopez Portillo before Reagan was inaugurated. Whether the U.S. will succeed in bringing Mexican oil under its control is problematical. The Mexican bourgeoisie, like significant sections of the Canadian, want to use their oil for their own purposes. The United States is thus faced with a struggle with both its neighbours over energy policy.

OTHERS

This is not the only potential area of conflict confronting the U.S. Another is South-East Asia, a potential source of non-OPEC oil. Part of the *rapprochement* between the United States and China involves oil. The Chinese government has granted a number of American companies the right to explore for oil in the South China Sea. Among these companies are Standard Oil (Indiana), ARCO, Mobil, Exxon, Texaco, Chevron, and Phillips. Much of the exploration is being conducted around Hainan Island, which is in territory claimed by Vietnam.[88] The question arises as to what Vietnam will do if oil is discovered and drilling rigs are set up. A conflict could arise between China and Vietnam over this area, which would carry with it a high potential for U.S. involvement. Furthermore, given the rising nationalism in China, there is a strong possibility that, if oil is discovered, the Chinese may well demand it be used for purposes other than those desired by the U.S. The U.S. may thus be involved in a struggle with China over oil.

No matter where the U.S. may go in the non-OPEC world, whether it be Canada, Mexico, China, Vietnam, Britain, Australia, Norway, or any other oil-bearing region, it will have difficulty in gaining control over world oil. Neither the United States nor its oil companies will ever be able to regain the dominance they enjoyed in the pre-OPEC days. Though the interests of all these non-OPEC oil regions differ in many respects, they have one interest in common: they do not want to see the re-emergence of an oil surplus controlled by the U.S. and its companies. This surplus, after all, could be used against them as well as against OPEC. Assuming that the U.S. succeeds in breaking up OPEC, then it will be in a position to play each oil country off against the other. Canada's demands could then be ignored just as easily as Ecuador's. Given this community of interests, it is highly unlikely that these countries will readily accede to U.S. dominance in oil.

### Industrialized Countries

There is also a continuing struggle between the U.S. and the other industrialized market economies, as well as among these economies themselves. This conflict involves the race for new energy technologies, technologies that will eventually replace oil as the motive force of industrial societies.

Though the predictions in recent years of an imminent oil shortage have been exaggerated, there is no doubt that the supply of oil is limited and that it will eventually be exhausted. Therefore society has to move from an oil to a non-oil era. In the more distant future, society will have to look increasingly for non-fossil fuel energy sources. The country that develops the new energy sources will be the one in the best position to assume a dominant role in the world. The major industrialized nations are therefore in a scramble to develop these new sources. The United States, to regain its dominance; the others, to gain power on their own.

This is why a coordinated energy policy is impossible. It is feasible only as a means of opposing OPEC's immediate demands. But even here unanimity has seldom, if ever, been achieved. France, for instance, has refused to join the IEA.

The war against OPEC moreover requires an energy surplus outside the control of OPEC. The building up of that surplus through non-OPEC oil and alternative energy sources takes time. In the short run a surplus can be achieved only through a reduction in demand for energy, and that reduction can come about in only two ways. The first is the more efficient use of energy, *i.e.*, what is normally considered to be conservation.[89] The second, and really the most effective, is a general reduction in the demand for all goods and services, *i.e.*, in a recession. The reappearance of an oil "surplus" in 1980 and the subsequent price cuts were directly related to the recession in the developed countries.

*Strength through Misery*

This downturn must be viewed as part of the economic warfare against OPEC. Kissinger actually spoke of achieving "strength through misery."[90] Levy noted that "a recession blamed on OPEC" might have other advantages. "It might," he pointed out, "gain support or acquiescence from other importing nations for American threats of economic or even military action against some Gulf Arab countries."[91] Since Levy made that statement, the U.S. has been building up a "rapid deployment" force for use in the Middle East but has, so far, achieved only limited success in obtaining cooperation from the other industrialized powers.[92] But World War II, it should be recalled, arose out of the depression of the 1930s. A severe recession in the 1980s could also lead to military confrontation.

The policy of "strength through misery" has a purpose in addition to the one of effectuating an immediate drop in the demand for energy. It is also necessary for the development of alternative energy sources, which requires not only advanced technology but also enormous amounts of capital. To raise these sums, national income has to be redistributed away from consumption and into investment. This is probably the main reason why the leading industrial nations have abandoned Keynesian remedies for the recession.

A Keynesian remedy implies an expansion of public sector spending as a means of raising consumption. This public sector spending can come only from potential investment funds. The thrust today, the so-called "supply-side economics," is to generate more investment funds through a reduction in public spending. It might be viewed as a capitalist version of Stalin's policy of rapid industrialization. This is the rationale behind the "anti-inflationary" and "austerity" programs carried out by all the market economies since the mid-1970s.

How long this economic warfare will last is hard to say. But as long as it does we can expect a continuous assault on living standards in all the industrialized countries. With this assault there will occur growing unemployment, labour unrest, civil strife, political instability, and increased tension among the major industrialized market powers. Whether the ultimate outcome will be the transformation of the economic conflict into a military one is impossible to predict.

One thing, however, seems clear. The end of the oil era spells the end of the age of American dominance. The U.S. can never regain its lost power. The American Empire, like all the empires of the past, will fade into oblivion. It remains to be seen whether this empire will end with a bang or a whimper.

**Notes**

1. P. Keenan, R. Dobias, M. O'Neil & N. Anderson, *Financial Analysis of a Group of Petroleum Companies* (Chase Manhattan Bank, New York, 1980), pp. 22–23.
2. P. Collier & D. Horowitz, *The Rockefellers: An American Dynasty* (New American Library, New York, 1976), p. 68.
3. Ibid., p. 54.
4. W. Greider & T. O'Toole, "Rockefeller Family Holdings Touch Every Economic Sphere," *Washington Post*, September 22, 1974.
5. Collier & Horowitz, *The Rockefellers*, p. 308.
6. U.S. House of Representatives, Committee on the Judiciary, *Hearings, Nomination of Nelson A. Rockefeller*, 93rd Congress, 2nd Session (U.S. Government Printing Office, Washington, 1974), pp. 847, 849.
7. Ibid., p. 845.
8. Ibid., pp. 847–48.

9. Excluded from the list were the holdings of the Rockefeller Foundation, the Rockefeller Brothers Fund, the Rockefeller Family Fund, and many charitable and educational foundations which have received shares in various companies from the Rockefeller family. It is problematical to what extent the investments of these foundations remain outside the control of the Rockefellers. Professors G. William Domhoff and Charles Schwartz, in their study of the Rockefellers, have noted that the data "strongly suggests an overlap of activities and individuals between the family's philanthropies and its business." (Ibid., p. 771)
10. *New York Times*, December 8, 1974.
11. U.S. House, *Hearings, Nomination of Nelson A. Rockefeller*, p. 848.
12. *The Washington Post*, September 26, 1974.
13. Greider & O'Toole, "Rockefeller Family Holdings Touch Every Economic Sphere."
14. Eastern Airlines provides a similar example. According to the *New York Times* (December 8, 1974), Laurance Rockefeller owned only 125,000 shares of Eastern's 19 million shares of common stock, or considerably less than one per cent of the common stock outstanding. Yet, with this relatively small portion of shares, he is alleged to be "highly influential" in the airline's affairs.
15. U.S. House, *Hearings, Nomination of Nelson A. Rockefeller*, p. 848.
16. Collier & Horowitz, *The Rockefellers*, pp. 260, 262.
17. Ibid., p. 263.
18. Ibid., p. 264.
19. Ibid.
20. Ibid., p. 263.
21. U.S. House, *Hearings, Nomination of Nelson A. Rockefeller*, p. 769.
22. Ibid.
23. U.S. Federal Trade Commission, *Concentration Levels and Trends in the Energy Sector of the U.S. Economy* (Federal Trade Commission, Washington, 1974), p. 1.
24. Ibid., p. 22
25. Ibid., p. 264.
26. Ibid., p. 1.
27. *Wall Street Journal*, May 5, 1973.
28. *American Metal Market*, August 8, 1973.
29. *Wall Street Journal*, October 10, 1974.
30. *Wall Street Journal*, June 2, 1975.
31. *The Financial Post Survey of Mines and Energy Resources 1980* (Maclean-Hunter, Toronto, 1980), p. 18.
32. *Globe and Mail*, March 14, 1981.
33. *American Metal Market*, November 1, 1977.
34. *Wall Street Journal*, November 8, 1973.
35. *Wall Street Journal*, February 10, 1978.
36. *Globe and Mail*, March 14, 1981.
37. U.S. Federal Trade Commission, *Concentration Levels and Trends*, p. 213.
38. *Globe and Mail*, March 14, 1981.
39. *Chemical Marketing Reporter*, May 24, 1976.

40. *The Financial Post Survey*, p. 213.
41. *Edmonton Journal*, October 13, 1979.
42. *Oil and Gas Journal*, October 17, 1977.
43. *Journal of Commerce*, June 18, 1980.
44. M. Tanzer, *The Race for Resources, Continuing Struggles over Minerals and Fuels* (Monthly Review Press, New York and London, 1980), p. 128.
45. *Wall Street Journal*, January 13, 1977.
46. *Chemical Marketing Reporter*, January 17, 1977.
47. *Wall Street Journal*, November 26, 1976.
48. *Financial Times*, April 7, 1978.
49. *Solar Engineering Magazine*, November 1979, p. 46.
50. *Home Auto*, August 1974, p. 10.
51. *Journal of Commerce*, November 19, 1976.
52. *Wall Street Journal*, October 30, 1973.
53. *Wall Street Journal*, August 25, 1975.
54. *Wall Street Journal*, March 20, 1974.
55. *Datamation*, October 1977, p. 5.
56. *Business Week*, October 8, 1979.
57. *Electronic News*, January 30, 1978.
58. *Computer World*, January 29, 1979.
59. *Business Week*, April 28, 1980.
60. *Publisher's Weekly*, November 12, 1978.
61. *Wall Street Journal*, February 7, 1976.
62. *New York Times*, August 19, 1976.
63. "The Energy Rip-Off," *Economic Notes*, vol. 49, no. 3 (1981), p. 1.
64. *Trialogue*, no. 25 (1980/1981), pp. 2, 17.
65. J. Sawhill, K. Oshima & H. Maull, *Energy: Managing the Transition* (The Trilateral Commission, New York, 1978), p. 1.
66. *New York Times*, July 27, 1969.
67. *Business Week*, October 25, 1969.
68. *Business Week*, April 14, 1962; *New York Times*, July 27, 1969, and *Edmonton Journal*, May 15, 1975.
69. *New York Times*, July 27, 1969.
70. *Business Week*, April 14, 1962.
71. M. Tanzer, *The Political Economy of Oil and the Underdeveloped Countries* (Beacon Press, Boston, 1969), pp. 133–34.
72. Ibid., p. 133.
73. M. Tanzer, "Oil Exploration Strategies: Alternatives for the Third World" in P. Nore & T. Turner (eds.), *Oil and Class Struggle* (Zed Press, London, 1980), pp. 89–98.
74. Ibid., pp. 93–94.
75. *Oil and Gas Journal*, July 1, 1968.
76. Ibid.
77. L. Turner, *Oil Companies in the International System* (The Royal Institute of International Affairs, London, 1978), pp. 179, 199.
78. *Business Week*, April 14, 1962.
79. *New York Times*, July 27, 1969.
80. Sawhill, Oshima & Maull, *Energy: Managing the Transition*, p. 73.

81. "Kissinger on Oil, Food, and Trade," *Business Week*, January 13, 1975, p. 75.
82. Ibid., p. 67.
83. Ibid.
84. "Oil Boom in Mexico May Tilt Economic Picture," *International Petroleum Encyclopedia*, 1980 (The Penn Well Publishing Company, Tulsa, Oklahoma, 1980), p. 437.
85. *Globe and Mail*, July 2, 1981.
86. "Oil Boom in Mexico May Tilt Economic Picture," *International Petroleum Encyclopedia*, p. 438.
87. *Globe and Mail*, June 29, 1981.
88. *The Boston Globe*, May 12, 1980.
89. Strictly speaking the more efficient use of energy is not synonymous with conservation. An increase in efficiency implies a higher output from a given input. Conservation, on the other hand, relates to the temporal allocation of a resource so as to maximize the present value of its income stream. Since the present value of income received in the distant future is low, an optimal conservation policy might dictate producing more in the present and less in the future. To the extent that an increase in efficiency might reduce present production and consumption, it might conflict with an optimal conservation policy.
90. W. J. Levy Consultants Corporation, "Oil on New Terms," New York, 1974, p. 21.
91. Ibid., p. 22.
92. "The Oil Crisis: Is There a Military Option?" *The Defense Monitor*, vol. 8, no. 11 (1979), pp. 1–7.

# Glossary of Acronyms

APQ - Average Program Quantity, a formula used by the participants in the Iranian consortium to allocate their worldwide production
ARAMCO - Arabian-American Oil Company
BP - British Petroleum
CFP - Compagnie Française des Pétroles
CIA - Central Intelligence Agency
DAPG - Deutsche-Amerikanische Petroleum Gessellschaft
ECA - Economic Cooperation Administration
FIRA - Foreign Investment Review Agency
FTC - Federal Trade Commission
GATT - General Agreement on Trade and Tariffs
GNP - Gross National Product
IBEC - International Basic Economic Corporation
IEA - International Energy Agency
IMF - International Monetary Fund
IPC - International Petroleum Company, a subsidiary of Imperial Oil operating in Latin America
IPC - Iraq Petroleum Company, the successor to the Turkish Petroleum Company (TPC)
IPL - Interprovincial Pipe Line
MSP - Minimum Safeguard Price
NEB - National Energy Board
NEDC - Near East Development Corporation
NEP - National Energy Program
NIOC - National Iranian Oil Company
NOP - National Oil Policy
OECD - Organisation for Economic Cooperation and Development
OEEC - Organisation for European Economic Cooperation
OPEC - Organisation of Petroleum Exporting Countries
PAA - Petroleum Administration Act
PGRT - Petroleum and Gas Revenue Tax
PIP - Petroleum Incentives Program
SDR - Special Drawing Rights
TPC - Turkish Petroleum Company
UNECE - United Nations Economic Commission for Europe

# Select Bibliography

Adelman, M. A. "Efficiency of Resource Use in Crude Petroleum." *The Southern Economic Journal,* vol. 31, 1964.

———. *The World Petroleum Market.* Baltimore: The Johns Hopkins University Press, 1972.

Aitken, H. *American Capital and Canadian Resources.* Cambridge, Mass.: Harvard University Press, 1961.

Akins, J. E. "The Oil Crisis: This Time the Wolf Is Here." *Foreign Affairs*, vol. 51, no. 3, 1973.

Alperovitz, G. *Atomic Diplomacy: Hiroshima and Potsdam, the Use of the Atomic Bomb and the American Confrontation with Soviet Power.* New York: Simon & Schuster, 1978.

Armand, L. *Some Aspects of the European Energy Problem, Suggestions for Collective Action.* Paris: Organisation for European Economic Cooperation, 1955.

Blair, J. *The Control of Oil.* New York: Pantheon, 1976.

Canada. Director of Investigation and Research, Combines Investigation Act. *The State of Competition in the Canadian Petroleum Industry.* 7 vols. Ottawa: Minister of Supply and Services Canada, 1981.

———. *An Energy Strategy for Canada: Policies for Self-Reliance.* Ottawa: Minister of Supply and Services Canada, 1976.

———. The Minister of Energy, Mines and Resources. *An Energy Policy for Canada, Phase 1—Analysis.* Ottawa: Information Canada, 1973.

———. *The National Energy Program.* Ottawa: Energy, Mines and Resources Canada, 1980.

———. *The National Energy Program, Update 1982.* Ottawa: Minister of Energy, Mines and Resources Canada, 1982.

———. *Report of the Royal Commission on Corporate Concentration.* Ottawa: Minister of Supply and Services, 1978.

———. Royal Commission on Energy. *Second Report to His Excellency the Governor General in Council.* Ottawa: 1959.

Collier, P. and D. Horowitz. *The Rockefellers: An American Dynasty.* New York: New American Library, 1976.

*Concentration Levels and Trends in the Energy Sector of the U.S. Economy.* Washington: U.S. Federal Trade Commission, 1974.

Crane, D. *Controlling Interest, The Canadian Gas and Oil Stakes.* Toronto: McClelland & Stewart, 1982.

de la Tramerye. *The World Struggle for Oil.* New York: Knopf, 1924.

Engler, R. *The Politics of Oil, A Study of Private Power and Democratic Directions.* New York: The Macmillan Company, 1961.

Evans, D. *The Politics of Energy, The Emergence of the Superstate.* Toronto: Macmillan of Canada, 1978.

Fanning, L. *The Shift to World Petroleum Power Away from the United States.* Pittsburgh: Gulf Oil Company, 1958.

Fleming, D. *The Cold War and Its Origins, 1917–1960.* London: Allen & Unwin, 1961.
Frank, H. "The Pricing of Middle East Crude Oil." Ph.D. thesis, Columbia University, 1961.
Fraser, H. *Diplomatic Protection of American Petroleum Interests in Mesopotamia, Netherlands, East Indies, and Mexico.* Washington: U.S. Government Printing Office, 1945.
Galbraith, J. K. *American Capitalism.* Sentry edition. Boston: Houghton Mifflin, 1962.
———. "Power and the Useful Economist." *American Economic Review,* vol. 63, no. 1, 1973.
Gols, A. "United States Foreign Oil Investments." Ph.D. thesis, University of Oregon, 1961.
Gray, E. *The Great Canadian Oil Patch.* Toronto: Maclean-Hunter, 1970.
———. *The Impact of Oil: The Development of Canada's Oil Resources.* Toronto: The Ryerson Press/Maclean-Hunter, 1969.
Halliday, F. *Iran Dictatorship and Development.* Middlesex, Eng.: Penguin Books, 1979.
Hartley, H. *Europe's Growing Needs for Energy—How Can They Be Met?* Paris: Organisation for European Economic Cooperation, 1956.
Hartshorn, J. E. *Oil Companies and Governments, An Account of the International Oil Industry in Its Political Environment.* London: Faber & Faber, 1962.
Hobson, C. K. *Export of Capital.* London: Constable, 1914.
Horowitz, D. *The Free World Colossus: A Critique of American Foreign Policy in the Cold War.* New York: Hill & Wang, 1965.
Hymer, S. and R. Rowthorn. "Multinational Corporations and International Oligopoly: The Non-American Challenge." In *The International Corporation,* edited by C. P. Kindleberger. Cambridge, Mass.: The M.I.T. Press, 1970.
Issawi, C. and M. Yeganeh. *The Economics of Middle Eastern Oil.* New York: Praeger, 1962.
Jacoby, N. *Multinational Oil, A Study in Industrial Dynamics.* New York: Macmillan, 1974.
Jensen, W. *Energy and the Economy of Nations.* Henley-on-Thames: G. T. Foulis, 1970.
Keenan, P., R. Dobias, M. O'Neil and N. Anderson. *Financial Analysis of a Group of Petroleum Companies.* New York: Chase Manhattan Bank, 1980.
Kenen, P. B. *Giant among Nations, Problems in United States Foreign Economic Policy.* Chicago: Rand McNally, 1960.
Kennedy, W. J., ed. *Secret History of the Oil Companies in the Middle East.* 2 vols. Salisbury, N.C.: Documentary Publications, 1979.
Kindleberger, C. P. *American Investment Abroad: Six Lectures on Direct Investment.* New Haven and London: Yale University Press, 1969.
Kolko, G. *The Politics of War: The World and United States Foreign Policy, 1943–45.* New York: Vintage Books, 1968.
———. *The Roots of American Foreign Policy: An Analysis of Power and Purpose.* Boston: Beacon Press, 1969.

Kolko, J. and G. Kolko. *The Limits of Power, The World and United States Foreign Policy.* New York: Harper & Row, 1972.

Lawrence, D. *The True Story of Woodrow Wilson.* New York: Doran, 1924.

Laxer, J. *Canada's Energy Crisis.* Toronto: James Lewis & Samuel, 1974.

Laxer, J. and A. Martin, eds. *The Big Tough Expensive Job: Imperial Oil and the Canadian Economy.* Toronto: Press Porcepic, 1976.

Lenin, V. I. *Imperialism: The Highest Stage of Capitalism.* New York: International, 1969.

Leontieff, W. "Theoretical Assumptions and Unobserved Facts." *American Economic Review,* vol. 61, no. 1, 1971.

Levitt, K. *The Silent Surrender, the Multinational Corporation in Canada.* Toronto: Macmillan of Canada, 1970.

Levy, W. J. "An Atlantic-Japanese Energy Policy." *Foreign Policy,* no. 11, 1973.

———. *The Search for Oil in Developing Countries: A Problem of Scarce Resources and Its Implications for State and Private Enterprise.* Washington: International Bank for Reconstruction and Development, 1961.

Lewis, C. *American's Stake in International Investment.* Washington: The Brookings Institution, 1938.

*Long-Term Cooperation Programme.* Paris: International Energy Agency, 1976.

Maddox, R. *The New Left and the Origins of the Cold War.* Princeton, N.J.: Princeton University Press, 1973.

Magdoff, H. *The Age of Imperialism, The Economics of U.S. Foreign Policy.* New York and London: Modern Reader Paperbacks, 1969.

Manes, R. A. "Import Quotas, Prices and Profits in the Oil Industry." *The Southern Economic Journal,* vol. 30, 1963.

Marx, K. and F. Engels. *The Communist Manifesto.* New York: International, 1980.

"Memorandum of Agreement between the Government of Canada and the Government of Alberta relating to Energy Pricing and Taxation." September 1, 1981.

Menderhausen, H. *Dollar Shortage and Oil Surplus in 1949–1950.* Princeton, N.J.: Princeton University Press, 1950.

Montague, G. *The Rise and Progress of the Standard Oil Company.* New York: Harper & Brothers, 1903.

Morray, J. *From Yalta to Disarmament: Cold War Debate.* New York: MR Press, 1961.

Nearing, S. and J. Freeman. *Dollar Diplomacy.* Reprint. New York and London: Monthly Review Press, 1966.

Noreng, O. "Friends or Fellow Travellers? The Relationship of non-OPEC Exporters with OPEC." *The Journal of Energy and Development,* vol. 4, no. 20, 1979.

O'Connor, H. *The Empire of Oil.* New York: Monthly Review Press, 1962.

———. *World Crisis in Oil.* New York: Monthly Review Press, 1962.

O'Connor, R. *The Oil Barons: Men of Greed and Grandeur.* Boston: Little, Brown & Co., 1971.

Odell, P. *Oil and World Power: Background to the Oil Crisis.* 3rd edition. Middlesex, Eng.: Penguin Books, 1974.

"The Oil Crisis: Is There a Military Option?" *The Defense Monitor.* vol. 8, no. 11, 1979.

Plotnick, A. *Petroleum: Canadian Markets and United States Foreign Trade Policy.* Seattle: University of Washington Press, 1964.

Pratt, L. *The Tar Sands: Syncrude and the Politics of Oil.* Edmonton: Hurtig, 1976.

Richards, J. and L. Pratt. *Prairie Capitalism: Power and Influence in the New West.* Toronto: McClelland & Stewart, 1979.

Robertson, N., ed. *Origins of the Saudi Arabian Oil Empire, Secret U.S. Documents.* Salisbury, N.C.: Documentary Publications, 1979.

Robinson, A. *Towards a New Energy Pattern in Europe.* Paris: Organisation for European Economic Cooperation, 1960.

Robinson, J. "The Second Crisis of Economic Theory." *American Economic Review,* vol. 62, no. 2, 1972.

Roseman, F. and B. Wilkinson. "Who Benefits? The Alberta Energy Price Increases." *The Canadian Forum,* vol. 53, no. 629–30, 1973.

Ryerson, S. *Unequal Union: Roots of Crisis in the Canadas, 1815–1873.* Toronto: Progress Books, 1973.

Sampson, A. *The Seven Sisters, The Greatest Oil Companies and the World They Shaped.* New York: The Viking Press, 1976.

Sawhill, J., K. Oshima and H. Maull. *Energy: Managing the Transition.* New York: The Trilateral Commission, 1978.

Scherer, F. M. *Industrial Market Structure and Economic Performance.* Chicago: Rand McNally, 1970.

Schumpeter, J. *Capitalism, Socialism, and Democracy.* New York and London: Harper & Brothers, 1947.

Scotland, W. "The Federal Government's View of the Canadian Energy Situation." *Canadian National Forum.* Canadian National Committee—World Energy Conference, 1974.

Shaffer, E. H. *The Oil Import Program of the United States, An Evaluation.* New York: Praeger, 1968.

Smith, P. *The Treasure Seekers: The Men Who Built Home Oil.* Toronto: Macmillan of Canada, 1978.

Solberg, C. *Oil Power.* New York: Mason/Charter, 1976.

Solo, R. "Gearing Military R & D to Economic Growth." In *The Economic Impact of the Cold War: Sources and Readings,* edited by J. L. Clayton. New York: Harcourt, Brace & World, 1970.

Stork, J. *Middle East Oil and the Energy Crisis.* New York and London: Monthly Review Press, 1975.

Sucre, L. A. "The Impact of Multinational Oil Companies on Venezuela: The Oil Companies in General and Gulf Oil in Particular." In *The Nation-State and Transnational Corporations in Conflict with Special Reference to Latin America*, edited by J. P. Gunneman. New York and London: Praeger, 1975.

Tanzer, M. "Oil Exploration Strategies: Alternatives for the Third World." In *Oil and Class Struggle*, edited by P. Nore and T. Turner. London: Zed Pres, 1980.

———. *The Political Economy of International Oil and the Underdeveloped Countries.* Boston: Beacon Press, 1969.

———. *The Race for Resources, Continuing Struggles over Minerals and Fuels.*

New York and London: Monthly Review Press, 1980.
Tarbell, I. *The History of the Standard Oil Company.* 2 vols. New York: The Macmillan Company, 1933.
Thompson, S. "Prorationing of Oil in Alberta and Some Economic Implications." M.A. thesis, University of Alberta, 1968.
Tugendhat, C. and A. Hamilton. *Oil, the Biggest Business.* London: Eyre Methuen, 1975.
Tupper, A. and G. Doern, eds. *Public Corporations and Public Policy in Canada.* Montreal: The Institute for Public Policy, 1981.
Turner, L. *Oil Companies in the International System.* London: The Royal Institute of International Affairs, 1978.
United Nations Economic Commission for Europe. *The Price of Oil in Western Europe.* Geneva: UNECE, 1955.
U.S. Armed Forces Information School. *The Army Almanac.* Washington: U.S. Government Printing Office, 1950.
U.S. Cabinet Task Force on Oil Import Control. *The Oil Import Question: A Report on the Relationship of Oil Imports to National Security.* Washington: U.S. Government Printing Office, 1970.
U.S. Federal Trade Commission. *The International Petroleum Cartel.* Washington: U.S. Government Printing Office, 1952.
U.S. Fuel Administration. *Final Report of the United States Fuel Administration, 1917–19.* Washington: U.S. Government Printing Office, 1921.
U.S. House of Representatives, Committee on the Judiciary. *Hearings, Nomination of Nelson A. Rockefeller.* Washington: U.S. Government Printing Office, 1974.
U.S. Senate, Subcommittee on Multinational Corporations of the Committee on Foreign Relations. *A Documentary History of the Petroleum Reserves Corporation, 1943–1944.* Washington: U.S. Government Printing Office, 1974.
Vallenilla, L. *Oil: The Making of a New Economic Order, Venezuelan Oil and* OPEC. New York: McGraw-Hill, 1975.
Walden, J. "The International Petroleum Cartel—Private Power and Public Interest." *Journal of Public Law,* vol. 2, no. 1, 1962.
Wilkins, M. *The Emergence of Multinational Enterprise: American Business Abroad from the Colonial Era to 1914.* Cambridge, Mass.: Harvard University Press, 1970.
———. *The Maturing of Multinational Enterprise: American Business Abroad from 1914 to 1970.* Cambridge, Mass.: Harvard University Press, 1974.
Williams, W. A. *The Roots of the Modern American Empire: A Study of the Growth and Shaping of Social Consciousness in a Marketplace Society.* New York: Random House, 1969.
———. *The Shaping of American Diplomacy: Readings and Documents in American Foreign Relations.* 2 vols. Chicago: Rand McNally, 1956.
———. *The Tragedy of American Diplomacy.* New York: Dell Press, 1962.
Williamson, H. F., R. Andreano, A. Daum and G. Klose. *The American Petroleum Industry, The Age of Energy, 1899–1959.* Evanston, Ill.: Northwestern University Press, 1963.
Williamson, H. F. and A. R. Daum. *The American Petroleum Industry, The Age of Illumination.* Evanston, Ill.: Northwestern University Press, 1959.

# Index